Biology of Forensically Important Invertebrates

Shyamasree Ghosh • Dhriti Banerjee

Biology of Forensically Important Invertebrates

Springer

Shyamasree Ghosh
School of Biological Sciences
National Institute of Science Education and Research (NISER), an OCC of Homi Bhabha National Institute (HBNI)
Bhubaneswar, Odisha, India

Dhriti Banerjee
Zoological Survey of India, Ministry of Environment, Forest and Climate Change
Kolkata, West Bengal, India

ISBN 978-981-99-5025-6 ISBN 978-981-99-5026-3 (eBook)
https://doi.org/10.1007/978-981-99-5026-3

This Springer imprint is published by the registered company Springer Nature Singapore Pte Ltd.
The registered company address is: 152 Beach Road, #21-01/04 Gateway East, Singapore 189721, Singapore

Paper in this product is recyclable.

Preface

The insect world is well known for its diversity contributing to the overall biodiversity. Due to their their variability in the genetical, physical, morphological, behavourial, functional aspects, they can cope in diverse environments and conditions. Insects are known to influence agriculture, human, resources with roles spanning widely as pests or carriers of diseases, to pollinators, scavengers, model organism for study and research to serving as food for other animals and to sense death. Thus their immensely wide biodiversity, potentially offers them importance in study of their biology, ecology and evolution. Forensic invertebrates include a large group of invertebrates that play an important role in forensic sciences. Forensic entomology in itself is a vast science encompassing all arthropods that are of forensic importance. They find importance in detection of crimes and thus finds importance in the judicial system in determining the time and place of death. Examination of insects collected from dead, decomposed bodies enables forensic entomologists to estimate the minimum Post-Mortem Interval (PMI) and also determines the cause of death.

It has become an important tool in criminal investigations. In the present scenario, the role of forensic odontologists is not confined to hard tissue examination alone. Increased instances of forensic odontologists being involved in criminal investigations, as part of the forensic team, have necessitated the need for an increase in awareness of emerging sciences like forensic entomology and its applications in forensics.

While we discuss briefly the invertebrates of forensic importance, we focus our attention in greater details on the forensic entomology and flies under Order Diptera that are of forensic importance.

The book highlights the importance of biology, taxonomy and biodiversity of flies under Order Diptera with forensic importance. We discuss our observation and research on forensic flies highlighting their morphological features, succession and biology. We also discuss the Cuticular HydroCarbons (CHC) and spectrometry-based studies reported from flies and larvae of forensic importance.

DNA barcoding has been an essential molecular biology tool developed recently towards understanding, identifying and grouping of the organisms and finds importance in molecular taxonomy-based studies on forensic flies. We discuss the developments in its recent status.

Not only do forensic flies find importance in determining human deaths, but they also serve as important indicators of animal cruelty and are of veterinary importance. We discuss the importance of forensically important flies in the veterinary world.

Although there are researches being carried out throughout the globe, in this domain, more trained technical manpower is needed to properly identify the flies which require a very specific skill. Our main idea behind the book is to introduce the new researchers, academicians and scholars in a didactic manner to the vast field of forensic entomology highlighting the areas, need and the scope of further research and inculcating the spirit of research within the scholars of today and tomorrow.

Unique Features

1. We discuss the biodiversity of forensically important invertebrates.
2. We highlight biodiversity of forensic flies from across the globe and India.
3. We highlight the morphological features of forensic flies.
4. We discuss recent updates on the Cuticular HydroCarbons (CHC) from forensic flies.
5. We discuss DNA bar coding and forensic entomology: where we stand today.
6. Importance of forensic flies in the veterinary world.

Bhubaneswar, Odisha, India Shyamasree Ghosh
Kolkata, West Bengal, India Dhriti Banerjee

Acknowledgement

We are thankful to the God Almighty for everything we have.

We acknowledge all the hard work of technicians and scientists who have relentlessly worked towards the progress of research in the field of forensic entomology.

We thank Director NISER, Bhubaneswar, Odisha, India, HBNI and Director ZSI, India for the support in the study.

We thank wholeheartedly the entire publication team of Springer Nature for the successful completion of the book.

We acknowledge, Dr. Garima Hore, Dr. Jayita Sengupta and Mrs Rashmi Bhattacherjee from ZSI, Kolkata and Prof Gautam Kumar Saha, Calcutta University for their contribution in the study.

Shyamasree and Dhriti

Contents

About the Authors

Shyamasree Ghosh (MSc, PhD) is currently working as the Scientific Officer (F) in the School of Biological Sciences, NISER, Bhubaneswar. She graduated from Presidency College under Calcutta University with Zoology, Chemistry and Physiology. She pursued an MSc in Biotechnology under Calcutta University and completed her PhD from the Indian Institute of Chemical Biology (ICB), CSIR Calcutta, working on glycobiology and acute childhood leukaemia (ALL). In her postdoctoral work, she worked on stem cells, invertebrate immune system and nanotechnology. She has also served as faculty in a postgraduate college affiliated to Bangalore University. She has worked extensively in the domains of zoology, immunology, glycobiology, sialo-biology, stem cells, computational biology and nanotechnology and authored and coauthored more than 70 articles in peer-reviewed international journals of repute, 10 books, several chapters and popular articles. She serves as the Associate Editor for *Frontiers in Immunology* and is an editorial board member of the *Journal of Applied Zoological Researches* and *Heliyon*. She is a life member of different prestigious scientific bodies. She has been awarded by different scientific bodies.

Dhriti Banerjee (MSc, PhD) is currently serving as the Scientist G and Director at Zoological Survey of India (ZSI) under the Ministry of Environment, Forest and Climate Change of the Government of India. She graduated from Presidency College under Calcutta University in 1991. She pursued an MSc in Zoology under Calcutta University and completed her PhD from Kalyani University in 1999. She received several awards and fellowships. She joined ZSI almost three decades ago and has been relentlessly working in the domain of Entomology and created history as the first woman Director of ZSI. She has more than 25 years of Scientific and Administrative experiences. Her research interest lies in taxonomy, forensic dipterology, climate modelling of flies (Diptera), tropic interaction, community relation in flies, and biology of pollinators and Dipterans. She has published more than 100 research articles in peer-reviewed international journals and authored or coauthored numerous books and book chapters. She was instrumental in setting up several databases for ZSI and has served as the scientist in charge and coordinator for IT, Digitization and Databasing.

1 Forensic Entomology

Abstract

Insects are a major contributor to human lives, both with their beneficial and harmful role. The last two decades have witnessed their increasing use in tracing absconding murderers for human and wild life. Forensic entomology has developed in the recent years as a fascinating branch of entomology. Forensically important invertebrates include *Diptera*, *Coleoptera*, and others. We discuss in this chapter insect succession and forensic entomology, developmental analysis of carrion insects and utility in forensic entomology, estimation of PMI, forensic entomotoxicological analysis, techniques used in study of forensic entomology, ultramorphological analysis of forensically important insects by scanning electron microscope and molecular identification techniques and maggot analysis, emerging technique and insects, and legal perspective.

Keywords

Diptera · Coleoptera · Maggot analysis · Forensic entomology

Abbreviations

ADD	Accumulated degree days
ADH	Accumulated degree hours
B. peregrine	*Boettcherisca peregrine*
HCIFS	Harris County Institute of forensic Sciences
L. sericata	*Lucilia sericata*
mtDNA	Mitochondrial DNA
PMI	Postmortem interval
PMImin	Postmortem interval

S. Ghosh, D. Banerjee, *Biology of Forensically Important Invertebrates*,
https://doi.org/10.1007/978-981-99-5026-3_1

SEM	Scanning electron microscopy
TOC	Time of colonization

1.1 Introduction

Forensic entomology, a field of forensic sciences, has become one of the very fascinating branches of science globally that deals with precise analyses of carrion entomofauna and cadaver remains to decipher the postmortem interval (PMI), or the minimum time elapsed since death (Amendt et al. 2004; Byrd and Allen 2001; Catts and Goff 1992, Fig. 1.1). Post-death, some adult insect can sense death, arrive on the dead body, survive on the juices, and complete a life cycle on the dead body through stages of eggs, larva, pupa, and adult (Figs. 1.2 and 1.3).

The major role of medico-criminal entomology is to impart detailed knowledge regarding estimations of the time, reason, manner, and location of the investigated death, with the aid of all the aspects that can be deduced from the study of carrion-associated entomofauna (Campobasso and Introna 2001), thus deciphering events of crime causing loss to human or animal life.

Forensic entomology encompasses three major domains, namely, (i) medicolegal forensic entomology, (ii) urban forensic entomology, and (iii) stored-product forensic entomology. Over the years, the discipline has chiefly procured relevance in the medico-criminal context (Byrd and Allen 2001). Urban entomology primarily deals with problems emanating from pest insects such as flies, termites, cockroaches, and others concerned with the human environment (Byrd and Allen 2001). On the other hand, stored-product forensic entomology relates to controversies over arthropods

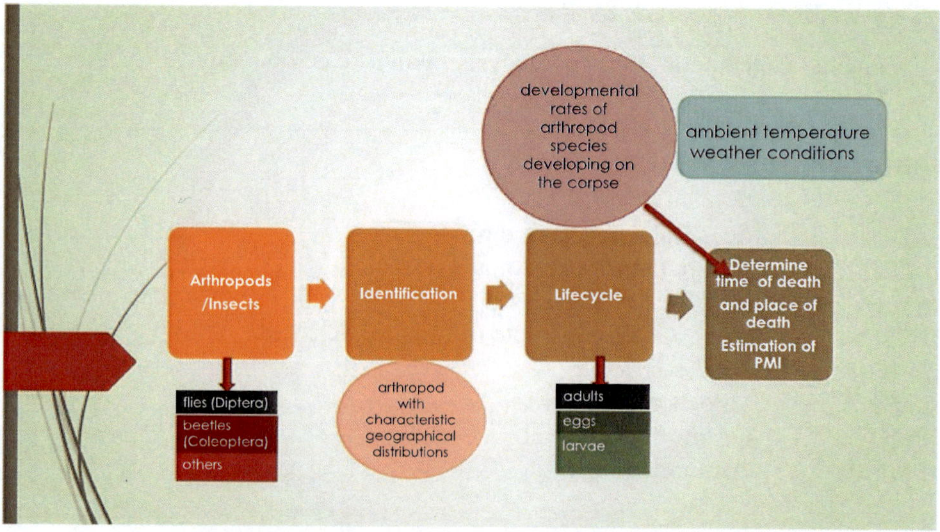

Fig. 1.1 Forensic entomology: a brief overview

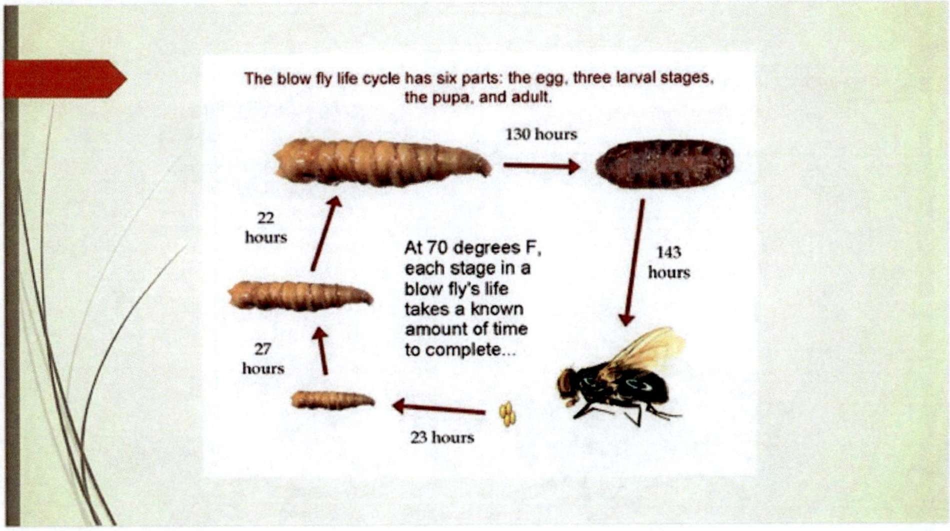

Fig. 1.2 Life cycle of a blow fly, Cleveland Museum of Natural History: Life cycle of the black blow fly
Female adult blow flies arrive on a cadaver within minutes to lay eggs, depositing around 250 eggs in the natural openings and wounds of the body. The eggs, within 24 h, hatch into first-stage maggots that feed and molt into second-stage maggots, which after several hours of feeding and growth molt into third-stage maggots. Masses of third-stage maggots producing heat can raise the temperature around them more than 10 ° C. After more feeding, the third-stage maggots metamorphize into adult flies. Image and legend adapted from NCBI (https://www.nlm.nih.gov/exhibition/visibleproofs/galleries/technologies/blowfly.html).

and arthropod parts in foodstuffs and related products (Byrd and Allen 2001). Forensic entomology, though generally holds implementations in human death investigations, nonetheless, has equal potential to unravel wildlife crimes (Anderson 1995; Byrd and Allen 2001).

Forensic pathologists are generally able to estimate the time elapsed during the first few hours after death, but when mortal remains are retrieved days, weeks, or even longer after death, general parameters like body temperature, and conditions like rigor mortis or livor mortis which forensic pathologists rely on, remain no longer suitable for determination of time of death (Amendt et al. 2004; Kashyap and Pillay 1989). In such cases, and after 72 h past death, insect faunal evidence are often the only accurate means for ascertaining the postmortem interval (PMI) (Amendt et al. 2004; Kashyap and Pillay 1989).

In addition to PMI estimations, critical examinations of the carrion-associated fauna are indeed worthwhile in determining whether the mortal remains have been relocated from an urban setup to a rural environment or vice versa (Catts and Haskell 1990; Charabidze et al. 2017; Erzinclioglu 1989). Furthermore, forensic entomotoxicology, a subdiscipline of forensic entomology, merges toxicological expertise into forensic entomology, in order to determine PMI, as well as reveal incidents of death, where toxicants and poisonous substances are presumably responsible for death (Bhardwaj et al. 2020).

Fig. 1.3 Developmental stages of forensic fly. Image and legend adapted from NCBI (https://www.nlm.nih.gov/exhibition/visibleproofs/galleries/technologies/blowfly.html). (**a**) Blow fly maggot, from Cleveland Museum of Natural History, (https://www.nlm.nih.gov/exhibition/visibleproofs/galleries/technologies/blowfly_image_2.html). (**b**) Blow fly puparium, the larvae becomes shorter and stouter with outer cuticle hardening into puparium and slowly darkens over 10 h and adults emerge in between 77 and 134 h after pupariation depending on the temperature. Image from NLM, Cleveland Museum of Natural History (https://www.nlm.nih.gov/exhibition/visibleproofs/galleries/technologies/blowfly_image_3.html). (**c**) Blow fly (*Phaenicia sericata*) Cleveland Museum of YPM ENT 599070. https://collections.peabody.yale.edu/search/Natural History. (**d**) Adult black blow fly and puparia (*Phormia regina*) Division of Entomology, Peabody Museum of Natural History, Yale University, (https://www.nlm.nih.gov/exhibition/visibleproofs/galleries/technologies/blowfly_image_5.html), specimen catalogued as YPM ENT 599070 reproduced with permission

1.2 Forensic Entomology: A Brief History

History of application of insects in deciphering crime dates back to the thirteenth century and reported from China (McKnight 1981), much earlier to even the coining of the term of forensic entomology (Amendt et al. 2004; Benecke 2001). The major advancement for this discipline was the first implementation of this field in a French courtroom, in 1850 (Bergeret 1855), where entomological information was affirmed as evidence for vindicating the current occupants of the residence from where the skeletonized remains of a child was located. Yovanovitch (1888) and Megnin (1894) were the first forensic investigators to assess insect succession on corpses and, thus, are accredited with properly establishing the science of forensic entomology (Amendt et al. 2004).

In recent times, forensic entomology has achieved recognition in many nations as a crucial forensic tool (Amendt et al. 2004; Greenberg 1991; Catts and Goff 1992;

Anderson 1995; Campobasso and Introna 2001). Extensive research conducted from various zoogeographic realms of the world is evident of the fact that forensic entomology has been a rapidly progressing field in forensic sciences.

A broad range of domains embrace forensic entomological research, globally, particularly extensive studies on developmental biology of forensic insects for PMI estimation (Anderson 2000; Greenberg 1991; Greenberg and Kunich 2000), studies on their ecology (Anderson 2001; Charabidze et al. 2014, 2017; Goff and Flynn 1991; Grassberger and Frank 2004; Smith 1986), entomotoxicological perspective (Campobasso et al. 2004; Goff et al. 1994a, b), DNA analysis in forensic entomology (Wells and Sperling 1999; Wells et al. 2001), and ultrastructural studies (Carriço et al. 2015; Sukontason et al. 2004, 2005, 2007, 2014, 2022).

A retrospective study on the forensic entomological research in India reveals its growth over the years as considerable amount of work has been conducted in the Indian states of Punjab, Himachal Pradesh, and Maharashtra as seen in the works of Singh and Greenberg (1994), Greenberg and Singh (1995), Singh and Bharti (2000, 2003), Bharti (2009, 2012), Bharti and Singh (2003), Sathe et al. (2013), Sathe and Jadhav (2013, 2014), Jadav and Sathe (2014), and Sharma et al. (2018), chiefly on inventorization of dipteran species colonizing carcasses at various geographical regions and ecological studies. Studies on filth-inhabiting flies in Kolkata, West Bengal, had been initiated by Joseph and Parui (1980); later an investigation on seasonal occurrence of *Calliphoridae* and *Sarcophagidae* was conducted with fish-baited traps in Calcutta by Das and Dasgupta (1982). Majumdar et al. (2008) assessed the synanthropy of carrion flies in 3 districts of Southern West Bengal, 12 localities in Midnapore, Hooghly, and North 24 Parganas. Diversity pattern of carrion flies using various parameters in three districts of West Bengal, India, has been studied by Majumdar et al. (2008). Comprehensive long-term study assessing seasonal variations in dipteran colonization and succession patterns on different carrion types in different habitats in and around Kolkata have been conducted by Hore et al. (2017a, b) and Hore (2021), and considerable work has also been conducted by Bhattacherjee et al. (2021) and Chakraborty et al. (2014a, b, c, 2015a, b, 2016a, b, c, 2017a, b) (Fig. 1.4).

1.3 Forensically Important Invertebrates

1.3.1 Diptera

The varied arthropod fauna constituting the carrion ecological community chiefly encompasses the groups Diptera (Fig.1.2), Coleoptera, Hymenoptera, and Arachnida (Amendt et al. 2004; Byrd and Allen 2001; Gruner et al. 2007). Although some of the carrion arthropods are lured directly by the carrion, which is utilized by them as a food source or an oviposition medium, yet other forensically relevant entomofauna are drawn to decomposing cadavers by the large assemblage of other insects, providing a food resource to them (Anderson 1995, 2000, 2001).

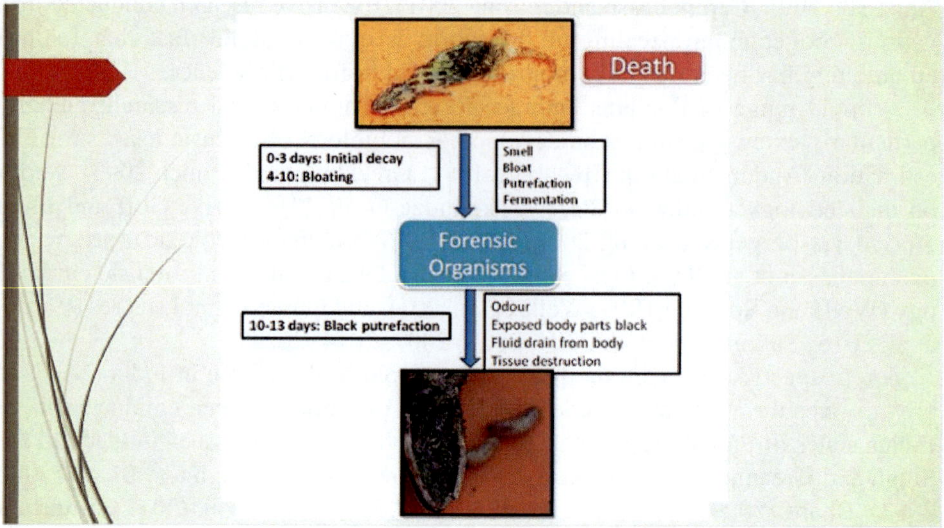

Fig. 1.4 Life cycle of *Sarcophaga (Liosarcophaga) dux on decomposing* dead gecko *Hemidactylus flaviviridis* Ruppel (1835 (Joseph and Parui 1980)). The decomposed stages of the dead gecko from fresh/initial decay (autolysis), putrefaction or bloating, and black putrefaction occur with the advent of forensic flies. Figure adapted from Chakraborty et al. (2014a, b, c) with permission

Necrophagous dipterans constitute the "first wave" of carrion entomofauna to commence the decomposition process, on accounts of their foraging behavior, and, thus, are often the first to colonize carcasses, generally within minutes after death. In fact, the adult dipterans feed upon the organic fluids oozing out of the corpse, and the dipteran larvae, commonly termed as maggots, on the other hand, are the key players of the decomposition process, secreting enzymes directly into the carrion, thus holds prime relevance with the liquefaction of the corpse tissues which further boosts the microbial activity (Campobasso and Introna 2001; Galante and Marcos-Garcia 2008).

Sarcophagidae, Calliphoridae, and Muscidae are proclaimed as predominant necrophagous dipteran families, globally, although a number of families like Anthomyiidae, Cecidomyiidae, Chloropidae, Drosophilidae, Dixidae, Ephydridae, Fanniidae, Phoridae, Piophilidae, Psychodidae, Milichiidae, Micropezidae, Neriidae, Sepsidae, Stratiomyidae, Sphaeroceridae, and Syrphidae colonizing carcasses have also been documented (Amendt et al. 2004; Byrd and Castner 2001a, b; Cavallari et al. 2014; Dupont et al. 2011; Farrell et al. 2015; Ivorra et al. 2021;Lindgren et al. 2015; Madeira-Ott et al. 2022; Martín-Vega 2011; Ndueze et al. 2013; Pereira de Sousa et al. 2014, 2016; Pereira de Sousa and Esposito 2015; Ren et al. 2018; Rochefort et al. 2015; Smith 1986; Song et al. 2008; Sukontason et al. 2022).

Taking into account the Indian context of forensic entomology, common necrophagous dipteran families reported to colonize corpses are primarily the Calliphoridae, Sarcophagidae, and Muscidae, whereas flies belonging to less prevalent families like Fanniidae, Phoridae, Piophilidae, Anthomyiidae, and Ulidiidae have also been documented (Archana et al. 2016; Bharti 2012; Bharti and Singh 2003; Bhattacherjee et al. 2021; Chakraborty et al. 2014a, b, c, 2015a, b, c, 2016a, b, c, 2017a, b; Hore et al. 2017a, b; Jadav and Sathe 2014; Joseph and Parui 1980; Khullar et al. 2014; Majumdar et al. 2007; Singh et al. 2016; Sinha 2009).

1.3.2 Coleoptera

Apart from dipterans, the second most forensically relevant entomofauna comprises of the Coleopterans (Bala and Singh 2015; Byrd and Castner 2001a, b; Kulshrestha and Satpathy 2001; Magni et al. 2015). Indeed, coleopteran beetles are proclaimed to reveal crucial entomological evidence in the medicolegal domain, especially concerning the dry human skeletal remains in the advanced stages of decomposition (Kulshrestha and Satpathy 2001).

The Coleopteran families of prime forensic relevance are Silphidae, Staphylinidae, Histeridae, Trogidae, Dermestidae, Cleridae, Nitidulidae, Carabidae, and Scarabaeidae (Almeida and Mise 2009; Bala and Singh 2015; Byrd and Castner 2001a, b; Farrell et al. 2015; Gennard 2012, Chap. 2).

Commonly termed as the carrion beetles, Silphidae constitutes a large coleopteran family, being almost cosmopolitan in distribution and comprising of more than 1500 species (Byrd and Castner 2001a, b).Interestingly, certain members of this family are termed as sexton or burying beetles, owing to the fact that some species bury their carrion or carcass food source (Byrd and Castner 2001a, b). As a matter of fact, studies proclaim that the burying of the food source may presumably be a means of getting rid of competition with other carrion-devouring entomofauna such as flies (Byrd and Castner 2001a, b). In both the burying beetle and the carrion beetle groups (genus *Silpha*) of the silphids, the larvae of both the burying beetles and the carrion beetles presumably thrive on carrion, while the adults chiefly forage on maggots (Byrd and Castner 2001a, b).

The Dermestidae (skin beetles) and Cleridae (bone beetles) are reported to be the predominant beetle types infesting exposed human remains and revealing crucial information in minimum postmortem interval (PMI) estimation (Byrd and Castner 2001a, b; Charabidze et al. 2014; Kulshrestha and Satpathy 2001). *Dermestes* species are apparently considered to be the only necrophagous entomofauna devouring decomposing remains, in some dry environments (Magni et al. 2015). Moreover, *Dermestes* species (both the immature stages and their adults counterparts), their remains (cast skins and fecal material), and their artifacts (pupal chambers) are often found associated with ancient remains such as mummies and fossils (Magni et al. 2015).The presence of dermestid beetles or their sawdust-like frass (fecal material) on cadavers is often a sign which reveals that quite some time has elapsed since death (Byrd and Castner 2001a, b).

Histeridae, generally known as clown beetles, are common predators that thrive primarily on dipteran larvae on carrion and dung (Szelecz et al. 2018).They are also reported to feed on the larvae of dermestid beetles (Byrd and Castner 2001a, b). Hide beetles (family Trogidae) and scarab beetles (family Scarabaeidae) are reported to be of considerable forensic importance, as well, as they are often found to be associated with carrion (Byrd and Castner 2001a, b).

1.3.3 Others

Although Dipterans and Coleopterans are the predominant arthropods colonizing carcasses, insects belonging to orders Hymenoptera, Lepidoptera, Mecoptera, and some members of Arachnida (order Araneae) and mites (families Acaridae, Lardoglyphidae, and Histiostomatidae) are also reported from a number of forensic entomological case studies, worldwide (Byrd and Castner 2001a, b; Kadej et al. 2020; Lindgren et al. 2015; O'Connor 2009; O'Connor et al. 2015; Ramos-Pastrana et al. 2018; Turner 2009).

1.4 Insect Succession and Forensic Entomology

Carrion constitutes a unique, transient microhabitat, marked primarily by its rapid degradation as a consequence of the waves of ecological succession by the arthropods that colonize it (Galante and Marcos-Garcia 2008). The odors arising from a corpse change as the process of decomposition progresses over time, becoming more alluring to some arthropod species and less attractive to others (Byrd and Castner 2001a, b).

Even though blow flies visit carcasses minutes after death, they are no longer drawn to the carrion after a certain stage of progression of the decomposition process, characterized by mummified or dry remains (Byrd and Castner 2001a, b; Nuorteva 1970, 1977).

Insect colonization of cadavers is dependent on a number of factors, yet one of utmost importance is the geographic region or bio-geoclimatic zone in which the remains are retrieved (Byrd and Castner 2001a, b). The bio-geoclimatic region delineates the habitat, vegetation, soil type, as well as the meteorological conditions of the area (Byrd and Castner 2001a, b).This seemingly has a profound impact on the types and species of insects present, including their seasonal availability (Byrd and Castner 2001a, b). The geoclimatic conditions affect the overall process of decomposition of the mortal remains, which in turn affects insect colonization (Byrd and Castner 2001a, b). Majority of families of carrion insects are comparatively wide-ranging, yet still the respective species associated in decomposition varies region-wise (Byrd and Castner 2001a, b).

The species associated with the sequential colonization of the remains as well as their arrival periods differs from one region to another. Perpetually, particular insect groups are reported to initiate carrion colonization, namely, blow flies and flesh flies (belonging to dipteran families Calliphoridae and Sarcophagidae, respectively), yet though the species involved will differ (Byrd and Castner 2001a, b).

Seasonal variations have a profound influence on the weather and the flora and fauna associated with a region. Hence, the entomo-faunal colonization of cadavers is also affected. Quite a number of blow fly species differ in their abundance depending on the seasonal changes. Besides the blow flies, majority of the other carrion insects are affected by seasonal variations and display particular peaks of activity (Byrd and Castner 2001a, b).

As necrophagous dipterans are key players mediating carcass decomposition and maintenance of homeostasis in the ecosystem, extensive research has been conducted to apprehend their ecological value (Amendt et al. 2004; Anderson 1995; Byrd and Castner 2001a, b; Edwards et al. 1970; Fiene et al. 2014; Galante and Marcos-Garcia 2008; Kitchell et al. 1979; Michaud et al. 2015; Lussenhop 1992; Vossbrinck et al. 1979).

Ecological studies on blow flies and flesh flies to understand their spatial as well as temporal patterns of occurrence has led to the hypothesis that blow flies respond quickly to changing environments, some reported to be common in habitats subjected to anthropogenic disturbances; in contrast, others are firmly connected with undisturbed or better-preserved habitats (Beltran et al. 2012; Dufek et al. 2016; Ferraz et al. 2014; Mariluis et al. 2008; Patitucci et al. 2011).

The information derived from such research has led to the understanding that blow fly abundance and occurrence amidst seasons as well as between different habitat types are vital in detecting forensically relevant species, from particular geographical regions, thus providing baseline information to support forensic entomological research on criminal cases (Gruner et al. 2007; Battan-Horenstein et al. 2010).

Ecological studies on necrophagous sarcophagids conducted abroad state sarcophagid species as indicators of urban or rural habitats in forensic entomological investigations (Fremdt et al. 2014), as these necrophagous dipterans are sensitive to anthropogenic impact, with habitat loss from human interference perceived as the predominant reason of decline in their biodiversity (Ceballos et al. 2017, Maxwell et al. 2016).

Such hypotheses have crucial implications in the domain of forensic entomology as elucidated in recent forensic entomological studies by Grassberger and Frank (2004) and Matuszewski et al. (2008, 2013) that certain particular indicator dipteran species can presumably play a role as forensic indicators for corpse relocation (Erzinclioglu 1989; Catts and Haskell 1990).

Beetles are the second most noteworthy carrion arthropods, other than dipterans from the forensic entomological perspective (Bala and Singh 2015; Byrd and Castner 2001a, b; Kulshrestha and Satpathy 2001; Magni et al. 2015; Singh and Bala 2019). Indeed, insights on the ecology of forensically relevant beetles could not only aid in PMI/PAI estimation (Matuszewski 2011, 2012; Souza et al. 2014; Kumara et al. 2009; Ridgeway et al. 2014), but also be used to validate mPMI estimates from dipteran data (Goff and Flynn 1991). The fact that beetles colonize carrion later than flies probably characterizes their specificity with a particular stage of decomposition (Varatharajan and Sen 2000) or as an adaptation to minimize their competition with flies (Smith 1986). Subsequently, ecological studies on forensically important beetles have gained attention worldwide as evident by the works of Ramos-Pastrana et al. (2018), Thakur and Kumari (2019), Horenstein and Linhares (2010), Kocarek (2003), Bajerlein et al. (2011), Bala and Kaur (2014), Singh and Bala (2019), Caballero and Leon-Cortes (2014), Correa et al. (2014), Dekeirsschieter et al. (2013), and Santos et al. (2013), to name a few.

It has long been perceived that insects associated with vertebrate carrion exhibit PMI-dependent processes (Byrd and Castner 2001a, b; Hall 1990). Critical analysis of the succession of arthropod species detected on and within a body (Schoenly and Reid 1987) is one such means of determination of PMI (Byrd and Castner 2001a, b; Goff 1993). In contrast to larval development, a succession model includes facts about the time elapsed since death, the arrival of a particular arthropod species, and the stage of decomposition (Byrd and Castner 2001a, b). Consequently, it can be used to evaluate both the minimum and maximum PMI (Byrd and Castner 2001a, b; Schoenly 1992).

1.5 Carrion Insects: Developmental Analysis and Forensic Entomology

Insects being poikilothermic organisms usually display differential developmental patterns dependent on the temperature at which they are exposed (Busvine 1980; Byrd and Castner 2001a, b). Needless to say, it is utmost crucial that in-depth analysis of developmental thresholds and stage-specific degree-day accumulations for all predominant forensic species be conducted to ensure accurate results in PMI estimations (Byrd and Castner 2001a, b; Nuorteva 1970, 1977). Interestingly, particular relevance of forensic application of insect development model relies mostly on maggot mass temperature (Byrd and Castner 2001a, b). The metabolic heat produced by maggot masses on dead tissue can be adequate in raising the temperature of the mass quite above the ambient temperatures (Byrd and Castner 2001a, b). The maggot mass temperature is generally considered a function of the size of the maggot mass, developmental stage, location on corpse (which consequently influences the relationship of mass temperature to ambient temperature), and ambient temperatures (Byrd and Castner 2001a, b; Catts and Goff 1992). A general strategy in this regard is measuring the maggot mass temperature and utilizing it as a constant in evaluating the developmental period (Byrd and Castner 2001a, b).

Studies on the bionomics of necrophagous dipteran species have direct implications in PMI estimation in forensic entomological research (Nuorteva 1970, 1977). Hence, a plethora of research has been conducted on the developmental analysis of necrophagous dipterans, worldwide. Investigations on the development rates of predominant forensically relevant necrophagous dipterans like Calliphorids and Sarcophagids had been conducted by Anderson (2000), Byrd and Allen (2001), Bunchu et al. (2012), Gomes et al. (2006), Grassberger and Reiter (2002), Nabity et al. (2007), Nassu et al. (2014), Kumara et al. (2013), Defilippo et al. (2013), Shiravi et al. (2011), Warren and Anderson (2013), Grassberger and Reiter (2002), Zuha et al. (2012), Byrd and Butler (1997), Wang et al. (2017a, b), and Wells and

Kurahashi (1994) to state a few. The larvae of Sarcophagid flies are probably the most useful insects for forensic entomological investigations as they are associated with corpses at active stages of decomposition, usually after 24 h of death (Oliveira-Costa et al. 2001; Wells et al. 2001), therefore are potential candidates for PMI estimation.

Taking into account the Indian perspective, considerable amount of such research has also been conducted in these lines (Bharti and Singh 2003; Bharti et al. 2007; Bharti 2009; Patil 2013; Abd-AlGalil and Zambare 2015; Bansode et al. 2016; Siddiki and Zambare 2017; Bhosale and Bhosale 2018; Verma and Paul 2016; Babu et al. 2017; Joseph and Parui 1980; Nandi 2002; Hore 2021; Chakraborty et al. 2014a, b, c, 2015a, b, 2015c, 2016a, b, c).

1.6 PMI Estimation

The time elapsed since death, or postmortem interval (PMI), is a matter of utmost relevance in investigations of homicides and other untimely deaths (Byrd and Castner 2001a, b).

Although there are varied means of estimating PMI from carrion insects (Sharma et al. 2015a, b) yet, the two widely practiced approaches involve the analysis of the sequential colonization of the carrion by successive waves of insects and other arthropods, and the second method is dependent on the development of immature flies that are deposited on the carrion soon after death (Goff 1993; Byrd and Castner 2001a, b; Verma and Paul 2013).

The utilization of either approach also relies upon factors like season, climate, location, and treatment of the corpse (Byrd and Castner 2001a, b; Verma and Paul 2013). One such development-dependent method of PMI estimation involves evaluation of the accumulated degree days or hours (ADD or ADH). ADH values represent a certain number of "energy hours" that are essential for the development of insect larvae. The degree day or hour concept is based on the assumption that the developmental rate is proportional to the temperature within a typical species-specific temperature range (Byrd and Castner 2001a, b; Sharma et al. 2015a, b).

Interestingly, the presence of foreign substances, such as drugs and toxins, in the decomposing body can have an impact on the growth rate of feeding maggots, and hence, such information can be of profound importance in PMI estimation (Goff et al. 1994a, b; Verma and Paul 2013). Numerous studies have shown that antemortem use of different drugs and toxins influence maggot development rate, consequently resulting into an erroneous PMI estimation based on insect development (Verma and Paul 2013). For instance, studies reveal that inaccurate results of up to 29 h can occur in PMI estimates with heroin containing tissues dependent on the development of the fly *Boettcherisca peregrina* (*B peregrina* Goff et al. 1991a, b; Verma and Paul 2013). Similar results are documented for methamphetamine and amitriptyline and errors of up to 24 h can occur in PMI estimates with heroin on *Lucilia sericata* (*L sericata* Verma and Paul 2013).

1.7 Forensic Entomotoxicology

Forensic entomotoxicology, a subdiscipline of entomotoxicology (Hodecek 2020), merges toxicology into forensic entomology to evaluate minimum postmortem interval (PMI_{min}) and instances of death where toxicants and other poisonous substances are the suspected cause of death (Bhardwaj et al. 2020; Byrd and Castner 2001a, b; Gosselin et al. 2011). Entomotoxicology also explores the impacts caused by drugs and toxins on arthropod development so as to aid forensic PMI estimates (Introna et al. 2001).

The use of necrophagous species in qualitative drug detection is well documented and usually accepted by forensic toxicologists (Gosselin et al. 2011; Kharbouche et al. 2008). Indeed, the implementation of forensic entomotoxicology in investigations of death caused by poisoning has achieved popularity, owing to the fact that insects serve as an alternative means of evidence when conventional toxicological samples like tissue, body fluids, and internal organs have degraded or are no longer available (Badenhorst and Villet 2018;Chophi et al. 2019; Goff et al. 1994a, b; Gosselin et al. 2011).

Insects are of forensic importance as they utilize the mortal remains as food source for their larvae (Chophi et al. 2019; Goff 2009a, b). Xenobiotics like drugs and other poisonous substances that are present in the body tissues enter the metabolic system of the larvae while they are actively devouring on the cadaveric tissue (Chophi et al. 2019). Moreover, these xenobiotics are transferred to other carrion-feeding arthropods that predate on the larvae by means of the food chain (Chophi et al. 2019; Introna et al. 2001; Pounder 1991). Insects (larvae) also serve as a suitable toxicological sample as they are present in ample numbers on the carrion and the puparial case remains unaltered and unaffected for a prolonged time period (Bourel et al. 2001; Chophi et al. 2019; Introna et al. 2001).

Techniques such as immunoassays (Hedouin et al. 1999; Introna Jr. et al. 1990; Kintz et al. 1990a, b), high-performance liquid chromatography (HPLC, Goff et al. 1993; Gunn et al. 2006; Kintz et al. 1990a, b; Sadler et al. 1995; Wolff et al. 2004), liquid chromatography-mass spectrometry (LCMS, Bushby et al. 2012; Goff et al. 1997; Kharbouche et al. 2008;Tracqui et al. 2004), and gas chromatography-mass spectrometry (GCMS, Campobasso et al. 2004; Definis-Gojanović et al. 2007; Liu et al. 2009; Manhoff et al. 1991) have been routinely used to ascertain the presence of xenobiotics from the cadaver, larvae, and insects (Chophi et al. 2019).

The presence of toxic substances has been reported to either augment or diminish the larval growth rate (Campobasso et al. 2004; Ferraz et al. 2014; Introna et al. 2001; O'Brien and Turner 2004; Tabor et al. 2005; Zou et al. 2013). Hence, erroneous PMI estimation can occur if the presence of toxic substances in tissues is not taken into consideration (Chophi et al. 2019).

1.8 Insect Collection and Preparation for Molecular Studies

Insects are collected to study their molecular biology and ultrastructure, and they aid in identification of the species, provide estimate of the life cycle and the PMI, and infer at phylogenetic relationships among insect taxa. Molecular analyses of immature forms enable proper inference of their adult forms and enable understanding of host-insect interactions (Gibb and Oseto 2020; Lehmann et al. 2012). Representative samples of both live and preserved samples need to be collected and hold importance in forensic investigations. The developmental stages of fly eggs, larvae (maggots), pupae, newly emerging adults, adult beetles, beetle larvae, ants, etc. are important specimens in forensic investigations. Sweep nets are used to collect the adult flies and can be collected directly in 80% ethyl alcohol or by using anesthesia, using an insect kill jar and then placed into 80% ethyl alcohol.

Blow flies (Calliphoridae), muscid flies (Muscidae), and flesh flies (Sarcophagidae) are searched for at the crime scene, in the natural openings of the body or wounds, clothes, and ground-body interface. Empty pupal cases and newly emerged blow flies could indicate completion of one life cycle. Predatory and necrophagous beetles arrive later. Insect types and species vary throughout decomposition of the dead body, due to altered availability of habitat and food resources in the crime scene. Although other organisms like spiders, earthworms, and pill bugs may inhabit a dead decomposing body, they do not provide information on PMI estimations.

Ambient temperature, microhabitat of the corpse, and maggot mass are important parameters for entomological analyses and need to be recorded to determine PMI.

Clean, contaminant-free (free of pollen or parasites) specimens enable better and accurate analysis of DNA or RNA extractions. Specimens are preserved using ethanol, freezing on dry ice followed by keeping in ultralow freezer, then in liquid nitrogen or chemicals like RNALater, which helps preserve the more fragile RNA materials. For molecular studies appropriate labeling of specimens is required.

1.9 Morphological Analysis by Microscopy

Accurate morphological identification of both adult and immature stages of insects associated with corpses is of paramount worth in the realm of forensic entomology. Light and scanning electron microscope (SEM) helps impart precise knowledge about the insect species examined, by enhanced visualization of the external morphology of insect specimens in subtle details, which is unobtainable by light microscopy. SEM has used extensively towards procuring species-specific ultrastructural diagnostic characters, information derived from which can provide rapid and precise morphological identification of a species from other such related ones. Thus, on this account, it has become imperative in ultramorphological characterization and, hence, has been extensively used in forensic entomology.

Ultrastructural characterization of adult and immature stages of necrophagous Diptera of forensic importance, belonging to mostly the predominant forensic families of Calliphoridae, Sarcophagidae, and Muscidae has been a booming domain of forensic entomology research, of late, aimed at enriching the anatomical database of forensic Diptera, thus aiding in precise morphological identification.

In-depth analysis of sensory organs like the antenna and mouthparts of forensically dominant Calliphorid species such as *Calliphora vicina, Chrysomya albiceps, Chrysomya megacephala, Chrysomya rufifacies, Chrysomya nigripes, Hemipyrellia ligurriens, Lucilia cuprina, Lucilia sericata,* and Muscid species like *Musca domestica, Ophyra chalcogaster, Ophyra aenescens (Wiedemann, 1830), Ophyra albuquerquei, Ophyra capensis (Wiedemann), Hydrotaea armipes*, and *Muscina stabulans* has been conducted by Carriço et al. (2015), Darılmaz et al. (2019), Hore et al. (2017a, b, 2018), Sukontason et al. (2004, 2007), and Wang et al. (2014). Comparably, ultrastructural studies on sensory organs of adult forensic sarcophagid species have received less attention, with studies conducted on species like *Sarcophaga (Liosarcophaga) dux, Peckia (Peckia) chrysostoma,* and *Sarcophaga tibialis* (Sukontason et al. 2004, 2014; Carriço et al. 2020; Pezzi et al. 2016).

Ultramorphological analysis of the immature stages of forensically relevant necrophagous Diptera has also been done extensively on calliphorid species such as *Chrysomya albiceps*, *Hemipyrellia ligurriens*, *Hemilucilia segmentaria (Fabricius, 1805), Lucilia cuprina (Wiedemann, 1830),* and *Lucilia sinensis* (Mendonça et al. 2010, 2014; Sanit et al. 2013; Thyssen and Linhares 2007; Sukontason et al. 2008); sarcophagid species like *Sarcophaga dux, Sarcophaga ruficornis, Peckia (Peckia) chrysostoma, Ravinia belforti*, and *Sarcophaga (Liopygia) cultellata* (da Silva-Xavier and de Carvalho Queiroz 2016; Buenaventura and Pape 2013; Leite and Lopes 2022; Liu and Greenberg 1989; Dahlem 1991; Sukontason et al. 2003; Ubero-Pascal and Arnaldos 2010; Ubero-Pascal et al. 2015a, b; Singh et al. 2012; Samerjai et al. 2014); and necrophagous muscids, namely, *Hydrotaea dentipes, H. similis, H. armipes, H. aenescens*, and *Hydrotaea capensis* by Grzywacz et al. (2017).

1.10 Molecular Identification

Precise identification of forensic entomofauna is an essential prerequisite for an accurate forensic entomological investigation. Forensically relevant insects are generally identified utilizing dichotomous keys, which depend on morphological features (Benecke and Seifert 1999; Greenberg and Singh 1995; Liu and Greenberg 1989; Gemmellaro et al. 2019). Morphological identifications can be problematic as local keys are often unavailable and can be difficult to use, especially when the immature stages are examined (Gemmellaro et al. 2019). Moreover, certain keys cannot be utilized for identification if specimens are damaged (Gemmellaro et al. 2019). On the contrary, molecular identification can be a better method of identifying forensically important insects, regardless of the life cycle stage dealt with or specimen completeness (Gemmellaro et al. 2019).

Implying DNA methods can not only supplement the traditional morphological methods but also presumably impart information on population substructures (Ames et al. 2006; Hillis and Moritz 1990; Sharma et al. 2015a, b). This identification can be conducted on any life cycle stage without further rearing as well as on dead or maintained forms, (Sharma et al. 2015a, b).

Of late, mitochondrial DNA has primarily been used as molecular markers for forensic species (Ames et al. 2006; Amendt et al. 2004; Byrd and Castner 2001a, b; Sharma et al. 2015a, b). Mitochondria are present in profuse numbers inside cells, and, consequently, mitochondrial DNA is present in a much higher exact replicate number than nuclear DNA (Butler 2001). Hence, it is the favored goal while dealing with incomplete or old tissue samples (Butler 2001; Malgorn and Coquoz 1999; Vincent and Vian 2000; Guo et al. 2010a, b).

In terms of both size and structure, the mtDNA of a calliphorid fly shows close resemblance to that of *Drosophila yakuba* (Burla) (Diptera: Drosophilidae), the first fly species for which the entire mitochondrial DNA (mtDNA) sequence was delineated (Byrd and Castner 2001a, b; Clary and Wolstenholme 1985). Therefore, there is a great deal of basic biological information available concerning fly mtDNA, which renders it easier to design PCR primers and in interpretation of the results of any study on a new fly species (Byrd and Castner 2001a, b).

Sperling et al. (1994) were the first to exemplify how mtDNA sequence data from (easy to identify) adult specimens of flies of forensic importance could be utilized to identify immature forms of the same species (Byrd and Castner 2001a, b). Using similar approaches, Wells and Sperling (1999) observed that calliphorid species that can be difficult to separate taxonomically, even in the adult stage, possess distinctly different mtDNA (Byrd and Castner 2001a, b).

The mitochondrial (mt) DNA investigation of forensically important flies focuses mainly on the cytochrome oxidase subunit one (COI, Guo et al. 2010a, b; Sperling et al. 1994; Wallman et al. 2005; Nelson et al. 2007; Sharma et al. 2015a, b; Meiklejohn et al. 2011). Furthermore, even partial sequences of this gene have been verified to have adequate discrimination power (Malgorn and Coquoz 1999; Vincent and Vian 2000; Guo et al. 2010a, b; Sharma et al. 2015a, b; Alessandrini et al. 2008), which makes it particularly important for forensic applications.

1.11 Maggot Analysis: An Emerging Technique

In order to attain more accuracy in criminology aspects related to forensic sciences, experts from different parts of the world have paved a newfangled way of molecular analysis of maggots for channeling out the modus operandi as well as the time, cause, and place of death of the victim. The victimized or dead animals provide beds for colonization of different types of insects at different time frames. However, a neat observation of a carcass tells a big story where the order Diptera stands out among other arthropods. As fly larvae voraciously feed on animal remains, decomposition hastens. Forensic experts chiefly consider the species composition of different forensic flies that invade decomposing remains to precisely evaluate the minimum postmortem interval since the time of death during a forensic investigation, and nowadays, with growing advantages of machine learning and unswerving in-depth analysis using real-time mass spectrometry, the calculation of lapse time since fly infestation has become much easier (Beyramysoltan et al. 2020).

Blow flies are not only the first-line colonizers of cadavers and carcasses, but they can also detect pungent volatile chemicals released from the dead bodies, from a considerable distance (Beyramysoltan et al. 2020). After colonization the blow flies lay eggs on the remains which later hatches into full-grown larvae through instars. The maggots readily feed on the rotten tissue; however, with time as the number of different forensic fly species as well as the mass of deposited eggs increases enormously, it sincerely becomes difficult for a forensic entomologist to differentiate the amassed larva of different species from each other (Beyramysoltan et al. 2020). Thus, many of the larvae are maintained under laboratory conditions to let them become adult for precise identification. Though it is a great opportunity to unveil the identity of those forensic Diptera, the method is truly time-consuming, and we cannot alleviate the chances of dead larva collection (Beyramysoltan et al. 2020). Thus, to optimize the result, experts nowadays use molecular analysis (DNA analysis) to extricate different species. Honestly speaking the sequences of the genome of many of these flies have not been deciphered by sequencing yet which adds a drawback to this method (Beyramysoltan et al. 2020).

The multispecies analysis method developed by Rabi A. Musah and colleagues executed direct analysis in real-time high-resolution mass spectrometry to extract information about the diverse mixtures of maggot species in a sample of interest at molecular level and was able to construct a new hierarchical conformal predictor for applications in data (Beyramysoltan et al. 2020). As a result, for the first time, it was possible to separate each maggot from a combination of up to six species (Beyramysoltan et al. 2020).

Furthermore, genetic analysis of larval gut contents can be used to determine what type food it has taken as meal, and also the DNA analysis of larval gut can be used in identifying the species of the concerned fly (Campobasso et al. 2005). In several different circumstances, the DNA extraction actually works in correlating a maggot with a corpse that it had colonized so as to assuage the chances of evidence dispute removal of remains from a suspected crime scene, an alternative food source is in close vicinity of the crime scene or the body, and a chain-of-evidence dispute (Campobasso et al. 2005). However, since maggot gut content analysis is a quite brand-new area of study, there are still many limitations of the technique (Campobasso et al. 2005).

1.12 Forensic Entomology Techniques and Applications

The major application of forensic entomology takes in the estimation of the minimum time since any suspicious death either by reckoning the stages of development of the oldest necrophagous insects or by assessing the species composition of diverse forensic insects that colonized the cadaver (Amendt et al. 2011). Apart from the morpho-taxonomic characterization, different types of toxicological and molecular techniques can also be used in determining the probable causes of death and can also disclose the identity of a victim, by correlating the last meal of a larvae with the suspect even in cases of corpse relocation (Amendt et al. 2011). Evidently speaking, there are some groups of forensic flies that not only invades dead animals but also can develop on living tissues producing myiasis (Amendt et al. 2011). During forensic criminal investigations, many layers of evidences are required to be presented in the court of law for paving the way of verdict; thus, collection, preservation, identification, and handling of the insect evidences by trained forensic experts are vastly recommended (Amendt et al. 2011).

Forensic entomological evidences in several homicide cases in terms of studying the development and feeding behavior of all maggots on a corpse sometimes become tedious (Zehner et al. 2004). Thus, a distinguished research contribution by a group of experts opted for human microsatellite DNA analysis after isolating it from the gut of necrophagous larvae that nourished on putrefied human cadaver involving about 4 months of lapse time since death (Zehner et al. 2004). The technique utilizing human STR profiles was found to be capable of tracing the association of a fly larvae to a particular corpse and also paved the way of identifying the host species such as pig by cytochrome b gene analysis (Zehner et al. 2004).

The application of forensic entomological evidences being limited to homicide investigations reducing the chances of tracing the trajectory of death from other different aspects since long (Sanford 2017). However, Harris County Institute of forensic Sciences (HCIFS) has provided the prospect of applying forensic entomology to classify diverse types of death, their circumstances, and location to minutely understand species-level trends across different geographical zones (Sanford 2017). In retrospect the medicolegal investigations of 203 cases linked to forensic entomology including different means of death classifications, stages of decomposition, and a huge number of crime scenes (deceased transported from the hospital with history of myiasis or sting allergy, outdoor scenes or indoor scenes) were meticulously analyzed (Sanford 2017). Abiotic factors such as temperature which fluctuated between indoor and outdoor scenes and nonfluctuating relative humidity were also considered during death scene investigations (Sanford 2017). Furthermore, most of the natural cases involve indoor scenes (43.3%), and most homicide cases were found in outdoor scenes (12.3%), whereas the rest manner of death includes both indoor and outdoor scenes during the course of the study (Sanford 2017). *Synthesiomyia nudiseta* under Diptera: Muscidae, *Lucilia cuprina* (Wiedemann) under Diptera, Calliphoridae, *Megaselia scalaris* Loew under Diptera: Phoridae, *Blaesoxipha plinthopyga* (Wiedemann) under Diptera: Sarcophagidae, and most of the Sarcophagidae (including *Blaesoxipha plinthopyga (Wiedemann)* as indicators present in the indoor scenes were confirmed by a Binomial test, whereas in case of outdoor scenes, *Lucilia eximia* (Wiedemann) under Diptera: Calliphoridae was found to be the only noteworthy indicator (Sanford 2017).

Very recent research regarding the use of small microRNAs as putative markers of immature blow fly, *Cochliomyia macellaria*, development including the estimation of age has come up to the limelight as it would be helpful in chronological assessments during investigations of death (Hjelmen et al. 2022). Blow fly being a promising candidate has a wide variety of roles in agriculture, ecology, and forensic entomology (Hjelmen et al. 2022). Thus, to decipher developmental markers of blow fly, RNA sequencing was performed throughout the developmental stages, and differentially expressed miRNAs were also taken into account (Hjelmen et al. 2022). Almost 217 miRNAs were identified among which 10 and 17 miRNAs were remarkably differentially expressed in larval and intrapuparial samples, respectively, whereas 28 miRNAs were differentially expressed among sexes (Hjelmen et al. 2022). Thus, the emerging application of RNA sequencing in detecting miRNAs for age markers in the calliphorid species *Cochliomyia macellaria* has opened up a new avenue of research for forensic experts and to utilize the same under an array of abiotic and biotic environments (Hjelmen et al. 2022).

Extraordinary research by Anthropology Research Facility (University of Tennessee) about atypical blow fly colonization where colonization of blow fly has shown overstated delay or recolonization of blow fly even after termination of the feeding phase on human remains are very rare to find cases in forensics (Owings et al. 2022).

The estimation of minimum postmortem interval is crucial in death investigations that involves arthropod evidence (Owings et al. 2022). However, inter-individual disparity in human decomposition often poses challenges in the way of gathering of forensic entomological evidence that lacks data alignment in retrospect or if it is unexpectedly different. Thus, in order to evaluate the minimum PMI, it is quite obvious to gather evidences where colonization of blow fly is markedly unique in terms of decomposition as a time of colonization (TOC) assessment is designed by the blow fly larvae.

1.13 Insects, Forensic Entomology, and Legal Perspective

Over the past two decades, there has been a steady progress in the realm of forensic entomology (Kotzé et al. 2021). The escalated involvement in this discipline has led to expanded research efforts, and applications have culminated in over 1100 research and review articles, and books, published since 2000 (Kotzé et al. 2021).

To state a few examples, forensic entomological specimens collected in Malaysia from human decedents during crime scene investigations during the time period 2005 till 2010 have been reviewed by Kavitha et al. (2013). A total of 80 cases were documented and 93 specimens were collected. *S. nudiseta* has been reported to be an uncommon species found only on human cadavers from indoors (Kavitha et al. 2013). Moreover, single-fly infestation was documented from a total of 75 cases (93.75%), whereas 5 cases (6.25%) had double-fly infestations (Kavitha et al. 2013). Conclusively, although ample numbers of fly species were found on human decedents, the most prevalent ones are still those of genus *Chrysomya* (Kavitha et al. 2013).

Apart from forensic flies, French forensic entomology laboratories during the time span of 1994–2013 have documented larder beetles (Coleoptera: Dermestidae) in 81 (7.5%) of 1093 human cadavers studied (Charabidze et al. 2014). Dipteran family Calliphoridae was recorded in 88% of the cases, whereas Hydrotaea and Piophilidae were observed 40% of the time (Charabidze et al. 2014). Interestingly, *Necrobia* spp. (Coleoptera: Cleridae) was observed in about 46% of the cases (Charabidze et al. 2014). Forensic case reports from Thailand from 2000 to 2006 have reported dipteran family Calliphoridae of prime importance, though families Muscidae, Sarcophagidae, Piophilidae, Phoridae, and Stratiomyidae were also collected from human remains (Sukontason et al. 2007). Larvae and adults of the beetle *Dermestes maculates* DeGeer (Coleoptera: Dermestidae) were also reported from some cases (Sukontason et al. 2007). Calliphorids belonging to genus *Chrysomya* were documented mostly, along with *Phormia regina*, whereas *Dermestes maculatus* and *Necrobia rufipes* are the reported Coleopteran species from five medicolegal forensic entomological case reports from Punjab, India (Sharma et al. 2018).

1.14 Discussion

In this chapter we have discussed in brief the important aspects of forensic entomology and the biology of the invertebrates, arthropods, and Insects that play major role in deciphering crime sciences and potentially play a role in decoding relation between victim, suspect, and crime (Campobasso et al. 2001; Catts 1992; Goff 1994; Hore and Banerjee 2017). The major challenges in this field of biology lies in accurate identification of the fly species to understand the time of death. While the origin of the science of forensic entomology and usage of insects in decoding crime scenes dates back to early thirteenth century, advances in this field have been growing with the advancement of the molecular biology techniques and microscopy techniques, and it has been possible to further refine the identification of fly species inhabiting the dead decomposing body and therefore provide a proper estimate of PMI to understand the time elapsed from death and the probable time of the crime. We discuss in subsequent chapters the important aspects in eth Biology of forensically important invertebrates with special focus of the forensically important Dipterans.

References

Abd-AlGalil FM, Zambare SP (2015) Effect of temperature on the development of calliphorid Fly of forensic importance, Chrysomya rufifacies (Macquart, 1842). IJAR 3:1099–1103

Alessandrini F, Mazzanti M, Onofri V, Turchi C, Tagliabracci A (2008) MtDNA analysis for genetic identification of forensically important insects. Forensic Sci Int Genet Suppl Ser 1:584–585

Almeida LM, Mise KM (2009) Diagnosis and key of the main families and species of south American coleoptera of forensic importance. Rev Brasil Entomol 53(2):227–244

Amendt J, Krettek R, Zehner R (2004) Forensic entomology. Naturwissenschaften 91(2):51–65

Amendt J, Richards CS, Campobasso CP, Zehner R, Hall MJ (2011) Forensic entomology: applications and limitations. Forensic Sci Med Pathol 7(4):379–392

Ames C, Turner B, Daniel B (2006) The use of mitochondrial cytochrome oxidase I gene (COI) to differentiate two UK blowfly species—Calliphora vicina and Calliphora vomitoria. Forensic Sci Int 164(2–3):179–182

Anderson GS (1995) The use of insects in death investigations: an analysis of cases in British Columbia over a five year period. Can Soc Foren Sci J 28(4):277–292

Anderson GS (2000) Minimum and maximum development rates of some forensically important Calliphoridae (Diptera). J Foren Sci 45(4):824–832

Anderson GS (2001) Forensic entomology in British Columbia: a brief history vol. J Entomol Soc Brit Columbia 98:127–136

Archana M, D'Souza PE, Ojha R, Jalali SK (2016) DNA barcoding of flies commonly prevalent in poultry farms of Bengaluru district. J Entomol Zool Stud 4(4):228–233

Babu S, Sharma H, Upadhyay S (2017) Studies on the larval growth of forensically important flesh fly Sarcophaga dux Thompson 1869 (Diptera: Sarcophagidae) under outdoor ambient temperatures from Central India. IJAR 3(7):366–370

Badenhorst R, Villet MH (2018) The uses of Chrysomya megacephala (Fabricius, 1794) (Diptera: Calliphoridae) in forensic entomology. Forensic Sci Res 3(1):2–15

Bajerlein D, Matuszewski S, Konwerski S (2011) Insect succession on carrion: seasonality, habitat (coleoptera: Histeridae) visiting pig carrion exposed in various forests (western Poland). Pol J Ecol 2011(59):787–797

Bala M, Kaur P (2014) Insect faunal succession on buried piece of pork in the state of Punjab (India): a preliminary study. J Forensic Res 5:252

Bala M, Singh N (2015) Beetles and forensic entomology: a comprehensive review. J Ent Res 39(4):293–302

Bansode SA, More VR, Zambare SP (2016) Fahd M Abd Algalil, effect of constant temperature (20 °C, 25 °C, 300 °C, 35 °C, 40 °C) on the development of the calliphorid fly of forensic importance, Chrysomya megacephala (Fabricus, 1794). J Entomol Zool Stud 4(3):193–197

Battán Horenstein M, LinHares AX, Rosso B, García MD (2010) Decomposition and dipteran succession on pig carrion in central Argentina: ecological aspects and their importance to forensic science. Med Vet Entomol 24:16–25

Beltran YTP, Segura NA, Bello FJ (2012) Synanthropy of Calliphoridae and Sarcophagidae (Diptera) in Bogot a Colombia. NeotropEntomol 41:237–242

Benecke M (2001) A brief history of forensic entomology. Forensic Sci Int 120(1–2):2–14

Benecke M, Seifert B (1999) ForensischeEntomologie am BeispieleinesTötungsdeliktes, Eine kombinierteSpuren- und Liegezeitanalyse. (forensic entomology in a high profile murder case: a combined stain, and post mortem interval analysis.). Arch Kriminol 204:52–60

Bergeret M (1855) Infanticide, momification naturelle du cadavre. Ann Hyg Publique Med Leg 4: 442–452

Beyramysoltan S, Ventura MI, Rosati JY et al (2020) Identification of the species constituents of maggot populations feeding on decomposing remains—facilitation of the determination of post mortem interval and time since tissue infestation through application of machine learning and direct analysis in real time-mass spectrometry. Anal Chem 92(7):5439–5446

Bhardwaj T, Sharma S, Dalal J, Verma K (2020) The implication of morphometrics and growth rate of dipteran flies in forensic entomotoxicology research: a review. Naturwissenschaften 107(6): 50

Bharti M (2009) Studies on life cycles of forensically important flies Calliphora vicina and Muscado mestica, nebulo at different temperatures. J Entomol Res 33:273–275

Bharti M (2012) Altitudinal diversity of forensically important blow flies collected from decaying carcasses in Himalaya. Open Foren Sci J 5:1–3

Bharti M, Singh D (2003) Insect faunal succession on decayed rabbit carcasses in Punjab. India J Foren Sci 48:1133–1143

Bharti M, Singh D, Sharma YP (2007) Effect of temperature on the development of forensically important blowfly, Chrysomya megacephala (Fabricius) (Diptera: Calliphoridae). Entomon 32: 149–151

Bhattacherjee R, Banerjee D, Ghosh S (2021) Chronology of appearance pattern and colonization of nutrient recycling forensic flies on house rat carcasses (rattus rattus) in urban and semiurban habitats of southern part of West Bengal. Int J Entomol Res 6(2):148–156

Bhosale PA, Bhosale YM (2018) Studies on time duration, life cycle stages of family (Calliphoridae) species: Calliphora vicina and Lucilia cuprina during summer season. Int J Zool Stud 3(2):144–146

Bourel B, Tournel G, Hedouin V, Deveaux M, Goff ML, Gosset D (2001) Morphine extraction in necrophagous insects remains for determining ante-mortem opiate intoxication. Forensic Sci Int 120:127–131

Buenaventura E, Pape T (2013) Revision of the new world genus Peckia Roniveau-Desvoidy (Diptera: Sarcophagidae). Zootaxa 3622:1–87

Bunchu N, Thaipakdee C, Vitta A, Sanit S, Sukontason K, Sukontason KL (2012) Morphology and developmental rate of the blow Fly, Hemipyrellia ligurriens (Diptera: Calliphoridae): forensic entomology applications. J Parasitol Res 2012:371243

Bushby SK, Thomas N, Priemel PA, Coulter CV, Rades T, Kieser JA (2012) Determination of methylphenidate in calliphorid larvae by liquid-liquid extraction and liquid chromatography mass spectrometry—forensic entomotoxicology using an in vivo rat brain model. J Pharm Biomed Anal 70:456–461

Busvine JR (1980) Insects and hygiene, the biology and control of insect pests of medical and domestic importance. Springer, New York, NY

Butler JM (2001) Forensic DNA typing: biology and technology behind STR markers. Academic Press, San Diego, CA

Byrd JH, Allen JC (2001) The development of the black blow fly, Phormia regina (Meigen). Forensic Sci Int 120(1–2):79–88

Byrd JH, Butler JF (1997) Effects of temperature on Chrysomya rufifacies (Diptera: Calliphoridae) development. Med Entomol 34:353–358

Byrd JH, Castner JL (2001a) Forensic entomology: the utility of arthropods in legal investigations. Taylor and Francis, Boca Raton

Byrd JH, Castner JL (2001b) Insects of forensic importance. In: Forensic entomology: the utility of arthropods in legal investigations, 2nd edn. CRC Press, Boca Raton, FL, pp 39–126

Caballero U, Leon-Cortes JL (2014) Beetle succession and diversity between clothed sun-exposed and shaded pig carrion in a tropical dry forest landscape in southern Mexico. Forensic Sci Int 245:143–150

Campobasso CP, Introna F (2001) The forensic entomologist in the context of the forensic pathologist's role. Forensic Sci Int 120(1–2):132–139

Campobasso CP, Di Vella G, Introna F (2001) Factors affecting decomposition and Diptera colonization. Forensic Sci Int 120(1–2):18–27

Campobasso CP, Gherardi M, Caligara M, Sironi L, Introna F (2004) Drug analysis in blowfly larvae and in human tissues: a comparative study. Int J Leg Med 118(4):210–214

Campobasso CP, Linville JG, Wells JD, Introna F (2005) Forensic genetic analysis of insect gut contents. Am J Forensic Med Pathol 26(2):161–165

Carriço C, Mendonça PM, Cortinhas LB, Mallet JR, Queiroz MM (2015) Ultrastructural studies of some character of Diptera (Muscidae) of forensically importance. Acta Trop 142:96–102

Carriço C, Caetano RL, Nascimento TM, Pinto ZT (2020) Ultrastructural studies of sensilla in one fly of forensic importance. EntomoBrasilis 13:e849

Catts EP (1992) Problems in estimating the post-mortem interval in death investigations. J Agric Entomol 9:245–255

Catts EP, Goff ML (1992) Forensic entomology in criminal investigations. Annu Rev Entomol 37: 253–272

Catts EP, Haskell NH (1990) Entomology and death: a procedural guide. Joyce's Print Shop Inc., Clemson, pp 1–182

Cavallari ML, Baltazar FN, de Carvalho EC, Munoz DR, Tolezano JE (2014) A modified shannon trap for use in forensic entomology. Adv Entomol 2(2):69–75

Ceballos G, Ehrlich PR, Dirzo R (2017) Biological annihilation via the ongoing sixth mass extinction signaled by vertebrate population losses and declines. Proc Natl Acad Sci U S A 114:E6089–E6096

Chakraborty A, Ansar W, Ghosh A, Banerjee D (2014a) The first report of the life cycle of Sarcophaga (L) dux on dead reptilian carcass: their application as forensic indicators. Sch Acad. J Biosci 2(11):731–739. ISSN 2347–9515.

Chakraborty A, Ansar W, Ghosh S, Banerjee D (2014b) The first report of the life cycle of Sarcophaga (L) dux on dead reptilian carcass: their application as forensic indicators. Sch Acad J Biosci 2(11):731–739

Chakraborty A, Naskar A, Hazra S, Maity A, Banerjee D (2014c) Indian black scavenger flies (Insecta: Diptera: Sepsidae). ZSI ePublication. https://doi.org/10.13140/RG.2.1.1771.4403

Chakraborty A, Ghosh S, Ansar W, Banerjee D (2015a) Developmental analysis of immature stages of Sarcophaga albiceps (Meigen,1820) on Gallus gallus carcass: their application as forensic indicators. IOSR J Agric Vet Sci 8(8, Ver II):79–89

Chakraborty A, Banerjee D, Ghosh S, Ansar W (2015b) Thermophilic pupal endoparasitoids: Brachymeria minuta (Hymenoptera: Chalicididae) on forensic indicator Sarcophaga (Parasarcophaga) albiceps. Prommalia III:84–95

Chakraborty A, Parui P, Banerjee D (2015c) Checklist of long legged fly: (Insecta: Diptera: Empidoidea: Dolichopodidae) of India. J Pharm Biol Sci 10(5):87–108

Chakraborty A, Naskar A, Parui P, Dhriti Banerjee D (2016a) Developmental variation of Indian thermophilic variety of scuttle Fly *Megaselia (Megaselia) scalaris* (Loew, 1866) (Diptera: Phoridae) on different substrates. Scientifica 2016:4257081. https://doi.org/10.1155/2016/4257081. 9 pages

Chakraborty A, Saha G, Banerjee D (2016b) Developmental analysis of two different variety of Indian thermophilic blow flies: Chrysomya megacephala (Fabricius, 1794) and Lucilia cuprina (Wiedemann, 1830) (Diptera: Calliphoridae) on dead Gallus gallus (Linnaeus, 1758). J Entomol Zool Stud 4(5):881–889

Chakraborty A, Saha GK, Banerjee D (2016c) Developmental variation of two different variety of Indian blow flies: Chrysomya megacephala (Fabricius, 1794) and Lucilia cuprina (Wiedemann, 1830) (Diptera: Calliphoridae) on dead Gallus gallus (Linnaeus,1758). J Entomol Zool Stud 4(5):881–889

Chakraborty A, Banerjee D, Saha KG (2017a) Biosystematic checklist of common myiasis causing calliphorids and Sarcophagids flies in India of veterinary importance. Agric Vet Sci 1(2): 116–134

Chakraborty A, Hora G, Parui P, Saha GK, Banerjee DA (2017b) Biosystematic species inventory of Indian Sarcophagidae (Insecta: Diptera: Sarcophagidae). J Entomol Zool Stud 5(1):465–473

Charabidze D, Colard T, Vincent B, Pasquerault T, Hedouin V (2014) Involvement of larder beetles (coleoptera: Dermestidae) on human cadavers: a review of 81 forensic cases. Int J Leg Med 128(6):1021–1030

Charabidze D, Gosselin M, Hedouin V (2017) Use of necrophagous insects as evidence of cadaver relocation: myth or reality? Peer J 5:e3506

Chophi R, Sharma S, Sharma S, Singh R (2019) Forensic entomotoxicology: current concepts, trends and challenges. J Forensic Leg Med 67:28–36

Clary DO, Wolstenholme DR (1985) The mitochondrial DNA molecule of drosophila yakuba: nucleotide sequence, gene organization and genetic code. J Mol Evol 22:252–271

Correa RC, Almeida LM, Moura MO (2014) Coleoptera associated with buried carrion: potential forensic importance and seasonal composition. J Med Entomol 51:1057–1066

da Silva-Xavier A, de Carvalho Queiroz MM (2016) Ultrastructure analysis of the immature stages of Ravinia belforti (Diptera: Sarcophagidae), a species of medical-veterinary and forensic importance, by scanning electron microscopy. Acta Trop 159:192–199

Dahlem GA (1991) Sarcophagidae (Oestroidea). In: Stehr FW (ed) Immature insects, vol 2. Kendall/Hunt Publishing Company, Dubuque, pp 871–873

Darılmaz MC, Karataş D, Taşatan H (2019) Ultrastructural studies on the antenna of Diptera of forensic importance. Mun Ent Zool 14(1):197–205

Das, S. K. & Dasgupta, B. 1982. Seasonal occurrence of blow flies and flesh flies in Calcutta. Oriental Ins., 16(1):119–128

Defilippo F, Bonilauri P, Dottori M (2013) Effect of temperature on six different developmental landmarks within the pupal stage of the forensically important blowfly Calliphora vicina (Robineau-Desvoidy) (Diptera: Calliphoridae). J Forensic Sci 58(6):1554–1557

Definis-Gojanović M, Sutlović D, Britvić D, Kokan B (2007) Drug analysis in necrophagous flies and human tissues. ArhHig Rada Toksikol 58(3):313–316

Dekeirsschieter J, Frederick C, Verheggen FJ, Drugmand D, Haubruge E (2013) Diversity of forensic rove beetles (coleoptera, Staphylinidae) associated with decaying pig carcass in a forest biotope. J Forensic Sci 58(4):1032–1040

Dufek MI, Oscherov EB, Damborsky MP, Mulieri PR (2016) Assessment of the abundance and diversity of Calliphoridae and Sarcophagidae (Diptera) in sites with different degrees of human impact in the Ibera wetlands (Argentina). J Med Entomol 53(4):827–835

Dupont FYF, Champlain DL, Cyrille AAJ, Fe ́lix BBC. (2011) A preliminary study of arthropod associated with carrion in Yaounde, Cameroon: a first step in forensic entomology in Central Africa. J Ecol Nat Environ 3:215–220

Edwards CA, Reichle DE, Crossley DA Jr (1970) The role of soil invertebrates in turnover of organic matter and nutrients. In: Analysis of temperate forest ecosystems, vol 1. Springer, Berlin, Heidelberg, New York, pp 147–172

Erzinclioglu YZ (1989) Entomology, zoology and forensic science: the need for expansion. Foren Sci Int 43(3):209–213

Farrell JF, Whittington AE, Zalucki MP (2015) A review of necrophagous insects colonising human and animal cadavers in south-East Queensland. Australia Foren Sci Int 257:149–154

Ferraz ACP, Dallavechia DL, Silva DC et al (2014) Effects of the antibiotics gentamicin on the postembryonic development of Chrysomya putoria (Diptera: calliphoridae). J Insect Sci 14(1): 1–5

Fiene JG, Sword GA, Van Laerhoven SL, Tarone AM (2014) The role of spatial aggregation in forensic entomology. J Med Entomol 51(1):1–9

Fremdt H, Amendt J, Zehner R (2014) Diapause-specific gene expression in Calliphora vicina (Diptera: Calliphoridae)—a useful diagnostic tool for forensic entomology. Int J Leg Med 128(6):1001–1011

Galante E, Marcos-Garcia MA (2008) Decomposer insects. In: Capinera JL (ed) Encyclopedia of entomology. Springer, Dordrecht. https://doi.org/10.1007/978-1-4020-6359-6_845

Gemmellaro MD, Hamilton GC, Ware JL (2019) Review of molecular identification techniques for forensically important diptera. J Med Entomol 56(4):887–902

Gennard D (2012) Forensic entomology – an introduction, 2nd edn. John Wiley & Sons Ltd., Publications, UK

Gibb TJ, Oseto C (2020) Insect collection and identification, 2nd edn. Elsevier

Goff ML (1993) Estimation of postmortem interval using arthropod development and successional patterns. Foren Sci Rev 5(2):81–94

Goff ML, Lord WD (1994) Entomotoxicology. A new area for forensic investigation. Am J Foren Med Pathol 15(1):51–57

Goff ML (2009a) Early post-mortem changes and stages of decomposition in exposed cadavers. Exp Appl Acarol 49(1–2):21–36

Goff ML (2009b) Foren entomology. In: Resh VH, Carde RT (eds) Encyclopedia of insects, 2nd edn. Academic Press, San Diego., pp 381–386

Goff ML, Flynn MM (1991) Determination of postmortem interval by arthropod succession: a case study from the Hawaiian islands. J Foren Sci 36:607–614

Goff ML, Brown WA, Hewadikaram KA, Omori AI (1991a) Effect of heroin in decomposing tissues on the development rate of Boettcherisca peregrina (Diptera: Sarcophagidae) and implications of this effect on estimation of postmortem intervals using arthropod development patterns. J Foren Sci 36:537–542

Goff ML, Brown WA, Hewadikaram KA, Omori AI (1991b) Effects of heroin in decomposing tissues on the development rate of *Boettcherisca peregrina* (Diptera: Sarcophagidae) and implications of this effect on estimations of postmortem intervals using arthropod development patterns. J Foren Sci 36:537–542

Goff ML, Brown WA, Omori AI, LaPointe DA (1993) Preliminary observations of the effects of amitriptyline in decomposing tissues on the development of Parasarcophaga ruficornis (Diptera: sarcophagidae) and implications of this effect to estimation of postmortem interval. J Foren Sci 38(2):316–322

Goff ML, Brown WA, Omori AI, LaPointe DA (1994a) Preliminary observations of the effects of phencyclidine in decomposing tissues on the development of Parasarcophaga ruficornis. J Foren Sci 39:123–128

Goff ML, Brown WA, Omori AI, LaPointe DA (1994b) Preliminary observations of the effects of phencyclidine in decomposing tissues on the development of Parasarcophaga ruficornis (Diptera: Sarcophagidae). J Foren Sci 39:123–128

Goff ML, Miller ML, Paulson JD, Lord WD, Richards E, Omori AI (1997) Effects of 3,4-methylenedioxymethamphetamine in decomposing tissues on the development of Parasarcophaga ruficornis (Diptera: Sarcophagidae) and detection of the drug in postmortem blood, liver tissue, larvae, and puparia. J Foren Sci 42(2):276–280

Gomes L, Godoy WA, Von Zuben CJ (2006) A review of postfeeding larval dispersal in blowflies: implications for forensic entomology. Naturwissenschaften 93(5):207–215

Gosselin M, Wille SM, Fernandez Mdel M, Di Fazio V, Samyn N, De Boeck G, Bourel B (2011) Entomotoxicology, experimental set-up and interpretation for forensic toxicologists. Foren Sci Int 208(1–3):1–9

Grassberger M, Frank C (2004) Initial study of arthropod succession on pig carrion in a central European urban habitat. J Med Entomol 41(3):511–523

Grassberger M, Reiter C (2002) Effect of temperature on development of the forensically important holarctic blow fly Protophormia terraenovae (Robineau-Desvoidy) (Diptera: Calliphoridae). Foren Sci Int 128(3):177–182

Greenberg B (1991) Flies as forensic indicators. J Med Entomol 28(5):565–577

Greenberg B, Kunich JC (2000) Entomology and the law: flies as forensic indictors. Cambridge University Press, Cambridge

Greenberg B, Singh D (1995) Species identification of calliphorid (Diptera) eggs. J Med Entomol 32:21–26

Gruner SV, Slone DH, Capinera JL (2007) Forensically important Calliphoridae (Diptera) associated with pig carrion in rural north-Central Florida. J Med Entomol 44:509–515

Grzywacz A, Hall MJR, Pape T et al (2017) Muscidae (Diptera) of forensic importance—an identification key to third instar larvae of the western Palaearctic region and a catalogue of the muscid carrion community. Int J Leg Med 131:855–866

Gunn JA, Shelley C, Lewis SW, Toop T, Archer M (2006) The determination of morphine in the larvae of Calliphora stygia using flow injection analysis and HPLC with chemiluminescence detection. J Anal Toxicol 30(8):519–523

Guo Y, Jifeng C, Xinghua W, Lingmei L, Qinlai L et al (2010a) Identification of forensically important sarcophagid flies (Diptera; Sarcophagidae) based on COI gene in China. Rom J Legal Med 18:217–224

Guo YD, Cai JF, Li X, Xiong F, Su RN et al (2010b) (2010b) identification of the forensically important sarcophagid flies Boettcherisca peregrina, Parasarcophaga albiceps and Parasarcophaga dux (Diptera; Sarcophagidae) based on COII gene in China. Trop Biomed 27: 451–460

Hall RD (1990) Medicocriminal entomology. In: Catts EP, Haskell NH (eds) Entomology and death: a procedural guide. Joyce's Print Shop, Clemson, pp 1–8

Hedouin GD, Bourel B, Martin-Bouyer L et al (1999) Morphine perfused rabbits: a tool for experiments in forensic entomotoxicology. J Foren Sci 44(2):347–350

Hillis DM, Moritz C (eds) (1990) Molecular systematics. Sinauer Associates, Sunderland

Hjelmen CE, Yuan Y, Parrott JJ, McGuane AS, Srivastav SP, Purcell AC, Pimsler ML, Sze SH, Tarone AM (2022) Identification and characterization of small RNA markers of age in the blow Fly *Cochliomyia macellaria* (Fabricius) (Diptera: Calliphoridae). Insects 13(10):948

Hodecek J (2020) Revisiting the concept of entomotoxicology. Foren Sci Int Synerg 2:282–286

Hore G (2021) Studies on necrophagous flies (Insecta: Diptera) of forensic importance in urban and semi-urban areas of Kolkata and adjoining areas. Ph.D. Thesis,. University of Calcutta, Kolkata, India

Hore G, Banerjee D (2017) Necrophagous flies (Insecta: Diptera) and their role in maintaining ecosystem balance. ENVIS newsletter (theme-ecosystem services). ENVIS Centre on Faunal Diver Zool Surv India 23(3):2–6

Hore G, Maity A, Naskar A, Ansar W, Ghosh S, Saha GK, Banerjee D (2017a) Scanning electron microscopic studies on antenna of Hemipyrellia ligurriens (Wiedemann, 1830) (Diptera: Calliphoridae): a blow fly species of forensic importance. Acta Trop 172:20–28

Hore G, Parui P, Saha GK, Banerjee D (2017b) Variations in colonization and succession pattern of dipteran flies of forensic importance on Indian mole-rat carcasses in urban and suburban localities of Kolkata, West Bengal: implications in corpse relocation studies. Malaysian J Med Res 1(3):52–62

Hore G, Saha GK, Banerjee D (2018) Sensory organs of forensically important fly Ophyra capensis (Wiedemann, 1818) (Diptera: Muscidae): a scanning electron microscopic study. Acta Trop 185:400–411

Horenstein BM, Linhares AX (2010) Seasonal composition and temporal succession of necrophagous and predator beetles on pig carrion in Central Argentina. Med Vet Entomol 25(4):395–401

Introna F Jr, Lo Dico C, Caplan YH, Smialek JE (1990) Opiate analysis in cadaveric blowfly larvae as an indicator of narcotic intoxication. J Foren Sci 35(1):118–122

Introna F, Campobasso CP, Goff ML (2001) Entomotoxicology. Foren Sci Int 120(1–2):42–47

Ivorra T, Martínez-Sánchez A, Rojo S (2021) Review of Synthesiomyia nudiseta (Diptera: Muscidae) as a useful tool in forensic entomology. Int J Leg Med 135(5):2003–2015

Jadav D, Sathe TV (2014) Altitudinal diversity of forensic blowflies (Diptera: Calliphoridae) of Western Ghats (Maharashtra). J Foren Res 5:6

Joseph ANT, Parui P (1980) Filth inhabiting flies (Diptera) of Calcutta city. Bullet Zool Surv India 3(1&2):1–12

Kadej M, Szleszkowski Ł, Thannhäuser A, Jurek T (2020) A mummified human corpse and associated insects of forensic importance in indoor conditions. Int J Leg Med 134(5):1963–1971

Kashyap VK, Pillay VV (1989) Efficacy of entomological method in estimation of postmortem interval: a comparative analysis. Foren Sci Int 40(3):245–250

Kavitha R, Nazni WA, Tan TC, Lee HL, Azirun MS (2013) Review of forensically important entomological specimens collected from human cadavers in Malaysia (2005–2010). J Foren Leg Med 20(5):480–482

Kharbouche H, Augsburger M, Cherix D et al (2008) Codeine accumulation and elimination in larvae, pupae, and imago of the blowfly Lucilia sericata and effects on its development. Int J Leg Med 122(3):205–211

Khullar N, Singh D (2014) 16S rRNA: a reliable mitochondrial gene for DNA amplificationfrom old and fragmented samples of forensically important blow flies (Diptera:Calliphoridae). World J Pharm Pharm Sci 3(10):1314–1319

Kintz P, Godelar B, Tracqui A, Mangin P, Lugnier A (1990a) Fly larvae: a new toxicological method of investigation in forensic medicine. J Foren Sci 35(1):204–207

Kintz P, Tracqui A, Ludes B et al (1990b) Fly larvae and their relevance in forensic toxicology. Am J Foren Med Pathol 11(1):63–65

Kitchell JF, O'Neill RV, Webb D, Gallep GW, Bartell SM, Koonce JF, Ausmus BS (1979) Consumer regulation of nutrient cycling. Bioscience 29:28–34

Kocarek P (2003) Decomposition and coleoptera succession on exposed carrion of small mammal in Opava, The Czech Republic. Eur J Soil Sci 39:31–45

Kotzé Z, Aimar S, Amendt J, Anderson GS, Bourguignon L, Hall MJR, Tomberlin JK (2021) The forensic entomology case report—a global perspective. Insects 12:283

Kulshrestha P, Satpathy DK (2001) Use of beetles in forensic entomology. Foren Sci Int 120(1–2): 15–17

Kumara TK, Hassan AA, Che Salmah MR, Bhupinder S (2009) The infestation of Dermestes ater (De Geer) on a human corpse in Malaysia. Trop Biomed 26(1):73–79

Kumara TK, Hassan AA, Salmah MR, Bhupinder S (2013) Larval growth of Liosarcophaga dux Thompson (Diptera: Sarcophagidae) under uncontrolled indoor temperatures in Malaysia. Southeast Asian J Trop Med Public Health 44(2):182–187

Lehmann P, Lyytinen A, Sinisalo T, Lindström L (2012) Population dependent effects of photoperiod on diapause related physiological traits in an invasive beetle (Leptinotarsa decemlineata). J Insect Physiol 58:1146–1158

Leite JD, Lopes RE (2022) Dissident genders and sexualities in the occupational therapy peer-reviewed literature: a scoping review. Am J Occup Ther 76(5):7605205160

Lindgren NK, Sisson MS, Archambeault AD, Rahlwes BC, Willett JR, Bucheli SR (2015) Four forensic entomology case studies: records and behavioral observations on seldom reported cadaver fauna with notes on relevant previous occurrences and ecology. J Med Entomol 52(2):143–150

Liu D, Greenberg B (1989) Immature stages of some flies of forensic importance. Ann Entomol Soc Am 82:80–93

Liu X, Shi Y, Wang H, Zhang RJ (2009) Determination of malathion levels and its effect on the development of *Chrysomya megacephala* (fabricius) in South China. Foren Sci Int 192:14–18

Lussenhop J (1992) Mechanisms of microarthropod microbial interactions in soil. Adv Ecol Res 23: 1–33

Madeira-Ott T, Souza CM, Bunde PRS, Ries ACR, Blochtein B, Thyssen PJ (2022) Forensically relevant flesh flies (Diptera, Sarcophagidae, Sarcophaginae) of southern Brazil. J Med Entomol 59(2):488–507

Magni PA, Voss SC, Testi R, Borrini M, Dadour IR (2015) A biological and procedural review of forensically significant dermestes species (coleoptera: Dermestidae). J Med Entomol 52(5): 755–769

Majumdar SK, Jana I, Misra KK (2007) Synanthropy of carrion flies in three districts of southern West Bengal, India, 2008. Int J Ecol Environ Sci 33(1):29–39

Majumdar, S., Chaki, K. K. & Misra, K. K. (2008). Diversity of carrion flies in relation to human habitation. Proceedings of the Zoological Society, 61(1&2), pp 51–66

Malgorn Y, Coquoz R (1999) DNA typing for identification of some species of Calliphoridae. An interest in forensic entomology. Foren Sci Int 102:111–119

Manhoff DT, Hood I, Caputo F, Perry J, Rosen S, Mirchandani HJ (1991) Cocaine in decomposed human remains. J Foren Sci 36(6):1732–1735

Mariluis JC, Schnack JA, Mulieri PR, Patitucci LD (2008) Calliphoridae (Diptera) de ambientes no habitados, suburbanos y urbanos en tres localidades del sudeste patagónico. Revista de la Soc Entomol Argentina 67(1–2):107–114

Martín-Vega D (2011) Skipping clues: forensic importance of the family Piophilidae (Diptera). Foren Sci Int 212(1–3):1–5

Matuszewski S (2011) Estimating the pre-appearance interval from temperature in Necrodes littoralis L. (coleoptera: Silphidae). Foren Sci Int 212(1–3):180–188

Matuszewski S (2012) Estimating the preappearance interval from temperature in Creophilus maxillosus L. (Coleoptera: Staphylinidae). J Foren Sci 57(1):136–145

Matuszewski S, Bajerlein D, Konwerski S, Szpila K (2008) An initial study of insect succession and carrion decomposition in various forest habitats of Central Europe. Foren Sci Int 180(2–3): 61–69

Matuszewski S, Szafałowicz M, Jarmusz M (2013) Insects colonising carcasses in open and forest habitats of Central Europe: search for indicators of corpse relocation. Foren Sci Int 231(1–3): 234–239

Maxwell SL, Fuller RA, Brooks TM, Watson JE (2016) Biodiversity: the ravages of guns, nets and bulldozers. Nature 536:143–145

McKnight BE (1981) The washing away of wrongs: forensic medicine in thirteenth-century China. University of Michigan, Ann Arbor

Megnin P (1894) La Faune des Cadavres. Encyclopédie Scientifiques des Aide Memoire||. Masson, Gauthier-Villars et Fils, Paris, p 214

Meiklejohn KA, Wallman JF, Dowton M (2011) DNA-based identification of forensically important Australian Sarcophagidae (Diptera). Int J Leg Med 125(1):27–32

Mendonça PM, Santos-Mallet JR, Queiroz MMC (2010) Ultramorphological characteristics of immature stages of Chrysomya albiceps (Wiedemann, 1819) (Diptera: Calliphoridae), a fly specie of forensic importance. Microsc Res Tech 73:779–784

Mendonça PM, Barbosa RR, Carriço C, Cortinhas LB (2014) Reis dos Santos-Mallet J, Maria de Carvalho Queiroz M, ultrastructure of immature stages of Lucilia cuprina (Diptera: Calliphoridae) using scanning electron microscopy. Acta Trop 136:123–128

Michaud JP, Schoenly KG, Moreau G (2015) Rewriting ecological succession history: did carrion ecologists get there first? Q Rev Biol 90(1):45–66

Nabity PD, Higley LG, Heng-Moss TM (2007) Light-induced variability in development of forensically important blow fly Phormia regina (Diptera: Calliphoridae). J Med Entomol 44(2):351–358

Nandi BC (2002) Fauna of India and the adjacent countries-Diptera: Sarcophagidae (vol. X), vol i–xxiv. Zoological Survey of India, pp 1–608

Nassu MP, Thyssen PJ, Linhares AX (2014) Developmental rate of immatures of two fly species of forensic importance: Sarcophaga (Liopygia) ruficornis and Microcerella halli (Diptera: Sarcophagidae). Parasitol Res 113(1):217–222

Ndueze OU, Noutcha MAE, Umeozor OC, Okiwelu SN (2013) Arthropods associated with wildlife carcasses in lowland rainforest, Rivers state, Nigeria. Eur Exp Biol 3(5):111–114

Nelson LA, Wallman JF, Dowton M (2007) Using COI barcodes to identify forensically and medically important blowflies. Med Vet Entomol 21:44–52

Nuorteva P (1970) Histerid beetles as predators of blowflies (Diptera: Calliphoridae) in Finland. Ann Zool Fenn 7:195–198

Nuorteva P (1977) Sarcosaprophagous insects as forensic indicators. In: Tedeschi CG, Eckert WG, Tedeschi LG (eds) Forensic medicine: a study in trauma and environmental hazards, vol II. W. B. Saunders Company, Philadelphia, London, Toronto, pp 1072–1095

O'Brien C, Turner B (2004) Impact of paracetamol on Calliphora vicina larval development. Int J Leg Med 118(4):188–189

O'Connor BM (2009) Astigmatid mites (Acari: Sarcoptiformes) of forensic interest. Exp Appl Acarol 49(1–2):125–133

O'Connor BM, Pimsler M, Owings CG, Tomberlin JK (2015) Redescription of Myianoetus muscarum (Acari: Histiostomatidae) associated with human remains in Texas, USA, with designation of a neotype from Western Europe. J Med Entomol 52(4):539–550

Oliveira-Costa J, Mello-Patiu CA, Lopes SM (2001) Influência de diferentes fatores na frequência de dípteros muscóides em cadáveres humanos no Rio de Janeiro. Bol Mus Nac Nova série (Zool) 470:1–10

Owings CG, McKee-Zech HS, Schwing ST, Bugajski KN, Davis MC, Steadman DW (2022) Not by the book: observations of delayed oviposition and re-colonization of human remains by blow flies. Insects 13(10):879

Patil AJ (2013) Use of Parasarcophaga species for forensic investigations during rainy season in Ahmednagar MS, India. Indian J Life Sci 3(1):87–90

Patitucci LD, Matiluis JC, Aballa FH (2011) A new species of the south American genus Arthurella Albuquerque(Diptera: Muscidae), with a key to species and new records. Zootaxa 2810:56–62

Pereira de Sousa JR, Carvalho-Filho FDS, Esposito MC (2015) Distribution and abundance of necrophagous flies (Diptera: Calliphoridae and Sarcophagidae) in Maranhao, Northeastern Brazil. J Insect Sci 15(1):70. https://doi.org/10.1093/jisesa/iev

Pereira de Sousa JR, Esposito MC, Carvalho-Filho FDS, Juen L (2014) The potential uses of Sarcosaprophagous flesh flies and blowflies for the evaluation of the regeneration and conservation of forest clearings: a case study in the Amazon Forest. J Insect Sci 14(1):215. https://doi.org/10.1093/jisesa/ieu077

Pereira de Sousa JR, Carvalho-Filho FDS, Juen L, Esposito MC (2016) Evaluating the effects of different vegetation types on necrophagous fly communities (Diptera: Calliphoridae; Sarcophagidae): implications for conservation. PLoS One 11(10):e0164826. https://doi.org/10.1371/journal.Pone.0164826

Pezzi M, Whitmore D, Chicca M, Semeraro B, Brighi F, Leis M (2016) Ultrastructural morphology of the antenna and maxillary palp of Sarcophaga tibialis (Diptera: Sarcophagidae). J Med Entomol 53:1–8

Pounder DJ (1991) Forensic entomo-toxicology. J Foren Sci Soc 31(4):469–472

Ramos-Pastrana Y, Virgüez-Díaz Y, Wolff M (2018) Insects of forensic importance associated to cadaveric decomposition in a rural area of the Andean Amazon, Caquetá, Colombia. Acta Amazon 48:126–136

Ren L, Shang Y, Chen W, Meng F, Cai J, Zhu G, Chen L, Wang Y, Deng J, Guo Y (2018) A brief review of forensically important flesh flies (Diptera: Sarcophagidae). Foren Sci Res 3(1):16–26

Ridgeway JA, Midgley JM, Collett IJ, Villet MH (2014) Advantages of using development models of the carrion beetles Thanatophilus micans (Fabricius) and T. mutilatus (Castelneau) (coleoptera: Silphidae) for estimating minimum post mortem intervals, verified with case data. Int J Leg Med 128(1):207–220

Rochefort S, Giroux M, Savage J, Wheeler TA (2015) Key to forensically important Piophilidae (Diptera) in the Nearctic region. Can J Arthropod Identif 27. https://doi.org/10.3752/cjai.2015.27

Sadler DW, Fuke C, Court F, Pounder DJ (1995) Drug accumulation and elimination in Calliphora vicina larvae. Foren Sci Int 71(3):191–197

Samerjai C, Sanit S, Sukontason K, KlongKlaew T, Kurahashi H, Tomberlin JK, Morakote N, Wannasan A, Sukontason KL (2014) Morphology of puparia of flesh flies in Thailand. Trop Biomed 31:351–361

Sanford MR (2017) Insects and associated arthropods analyzed during medicolegal death investigations in Harris County, Texas, USA: January 2013–April 2016. PLoS One 12(6):e0179404

Sanit S, Sribanditmongkol P, Sukontason KL, Moophayak K, Klong-Klaew T, Yasanga T, Sukontason K (2013) Morphology and identification of fly eggs: application in forensic entomology. Trop Biomed 30:325–337

Santos WE, Alves ACF, Farias RCAP, Creao- Duarte AJ (2013) Ecological role of coleoptera associated with carcasses in caatinga. Entomo Brasilis 6:248–250

Sathe TV, Jadhav D (2013) Sarcophagid fauna (Diptera: Sarcophagidae) of Kolhapur district. J Adv Zool 34(85–87):21

Sathe TV, Jadhav D (2014) Ecology and control of eye flies, Siphunculina spp. (Chloropidae: Diptera). Int Nat J Pharma Biol Sci 5:214–220

Sathe TV, Sathe A, Sathe NT (2013) Diversity of dipterous forensic insects from Western Maharashtra, India. Int J Pharma Bio Sci 4:173–179

Schoenly K (1992) A statistical analysis of successional pat-terns in carrion-arthropod assemblages: implications for forensic entomology and determination of the PMI. J Foren Sci 37:1489–1513

Schoenly K, Reid W (1987) Dynamics of heterotrophic succession in carrion arthropod assemblages: discrete series or a continuum of change? Oecologia (Berl) 73:192–202

Sharma M, Singh D, Sharma AK (2015a) Mitochondrial DNA based identification of forensically important Indian flesh flies (Diptera: Sarcophagidae). Foren Sci Int 247:1–6

Sharma R, Garg RK, Gaur JR (2015b) Various methods for the estimation of the postmortem interval from Calliphoridae: a review. Egypt J Foren Sci 5:1–12

Sharma A, Bala M, Singh N (2018) Five case studies associated with forensically important entomofauna recovered from human corpses from Punjab, India. Foren Sci Criminal Inves 7(5):555721. https://doi.org/10.19080/JFSCI.2018.07.555721

Shiravi A, Mostafavi R, Akbarzadeh K, Oshaghi M (2011) Temperature requirements of some common forensically important blow and flesh flies (Diptera) under laboratory conditions. Iran J Arthropod Borne Dis 5(1):54–62

Siddiki S, Zambara SP (2017) Effect of seasonal temperature variation on the duration of life cycle stages of the fly of forensic importance, Parasarcophga dux (Thomson) (Diptera: Sarcophagidae). Int J Adv Res 5(11):265–269

Singh N, Bala M (2019) Succession study on forensically important coleoptera from India: a preliminary study and its forensic implications. Egypt J Foren Sci 9:66

Singh D, Bharti M (2000) Forensically important blow flies (Diptera: Calliphoridae) of Punjab (India). Uttar Pradesh J Zool 20(3):249–251

Singh D, Greenberg B (1994) Survival after submergence in the pupae of five species of blowflies (Diptera: Calliphoridae). J Med Entomol 31(5):757–759

Singh D, Garg R, Wadhawan B (2012) Ultramorphological characteristics of immature stages of a forensically important fly *Parasarcophaga ruficornis* (Fabricius)(Diptera: Sarcophagidae). Parasitol Res 110:821–831

Singh N, Kumari V, Singh NP (2016) An analysis of the ecological succession pattern of Diptera on the carcass of laboratory bred rats. Int J Agric Innov Res 4(5):2319–1473

Sinha SK (2009) Sarcophagidae, Calliphoridae and Muscidae (Diptera) of the Sundarbans biosphere reserve, West Bengal, India. Records of the Zoological Survey of India, Occasional Paper No. 308, pp. 1–46.

Smith KG (1986) A manual of forensic entomology. The Trustees of British Museum, London

Song ZK, Wang XW, Liang GQ (2008) Species identification of some common necrophagous flies in Guangdong province, southern China based on the rDNA internal transcribed spacer 2 (ITS2). Foren Sci Int 175:17–22

Souza ER, Rafael JA, Filho FFX, Da-Silva-Freitas JO, Oliveira Costa J, Ururahy RA (2014) First medicolegal forensic entomology case of Central Amazon: a suicide by hanging with incomplete suspension. EntomoBrasilis 7:12–15

Sperling FAH, Anderson GS, Hickey DA (1994) A DNA-based approach to the identification of insect species used for postmortem interval estimation. J Foren Sci 39:418–427

Sukontason KL, Sukontason K, Piangjai S, Boonchu N, Chaiwong T, Vogtsberger RC, Kuntalue B, Thijuk N, Olson JK (2003) Larval morphology of Chrysomya megacephala (Fabricius) (Diptera: Calliphoridae) using scanning electron microscopy. J Vector Ecol 28(1):47–52

Sukontason K, Sukontason KL, Piangjai S, Boonchu N, Chaiwong T, Ngern-Klun R, Sripakdee D, Vogtsberger RC, Olson JK (2004) Antennal sensilla of some forensically important flies in families Calliphoridae, Sarcophagidae Muscidae. Micron 35(8):671–679

Sukontason K, Sukontason KL, Piangjai S, Choochote W, Vogtsberger RC (2005) Ultrastructure of the ommatrichia in Megaselia scalaris (Loew) (Diptera: Phoridae). Micron 36(2):191–194

Sukontason K, Narongchai P, Kanchai C, Vichairat K, Sribanditmongkol P, Bhoopat T, Kurahashi H, Chockjamsai M, Piangjai S, Bunchu N, Vongvivach S, Samai W, Chaiwong T, Methanitikorn R, Ngern-Klun R, Sripakdee D, Boonsriwong W, Siriwattanarungsee S, Srimuangwong C, Hanterdsith B, Chaiwan K, Srisuwan C, Upakut S, Moopayak K, Vogtsberger RC, Olson JK, Sukontason KL (2007) Forensic entomology cases in Thailand: a review of cases from 2000 to 2006. Parasitol Res 101(5):1417–1423

Sukontason K, Piangjai S, Siriwattanarungsee S, Sukontason KL (2008) Morphology and developmental rate of blowflies Chrysomya megacephala and Chrysomya rufifacies in Thailand: application in forensic entomology. Parasitol Res 102(6):1207–1216

Sukontason KL, Sanit S, Klong-Klaew T, Tomberlin JK, Sukontason K (2014) Sarcophaga (Liosarcophaga) dux (Diptera: Sarcophagidae): a flesh fly species of medical importance. Biol Res 47(1):14

Sukontason KL, Sanit S, Limsopatham K, Wannasan A, Somboon P, Sukontason K (2022) Chrysomya pinguis (Walker) (Diptera: Calliphoridae), blow fly of forensic importance: a review of bionomics and forensic entomology appraisal. Acta Trop 232:106506

Szelecz I, Feddern N, Seppey CVW, Amendt J, Mitchell EAD (2018) The importance of Saprinus semistriatus (coleoptera: Histeridae) for estimating the minimum post-mortem interval. Leg Med (Tokyo) 30:21–27

Tabor KL, Fell RD, Brewster CC, Pelzer K, Behonick GS (2005) Effect of antemortem ingestion of ethanol on insect succession patterns and development of Phormia regina (Diptera: calliphoridae). J Med Entomol 42(3):481–489

Thakur J, Kumari U (2019) Studies on incidence and succession of forensically important insects associated with decomposition of goat flesh. J Entomol Zool Stud 7(4):1427–1431

Thyssen PJ, Linhares AX (2007) First description of the immature stages of Hemilucilia segmentaria (Diptera: Calliphoridae). Biol Res 40(3):271–280. Epub 2008 Apr 17

Tracqui A, Keyser-Tracqui C, Kintz P, Ludes B (2004) Entomotoxicology for the forensic toxicologist: much ado about nothing? Int J Leg Med 118(4):194–196

Turner B (2009) Forensic entomology: a template for forensic acarology? Exp Appl Acarol 49(1–2):15–20

Ubero-Pascal N, Arnaldos RI (2010) López-Esclapez and García MD. Microscopy and forensic entomology. In: Méndez-Vilas A, Díaz J (eds) Microscopy: science, technology, applications and education. Formatex Research Center, Badajoz

Ubero-Pascal N, Paños Á, García MD, Presa JJ, Torres B, Arnaldos MI (2015a) Micromorphology of immature stages of Sarcophaga (Liopygia) cultellata Pandellé, 1896 (Diptera: Sarcophagidae), a forensically important fly. Microsc Res Tech 78(2):148–172

Ubero-Pascal N, Paños Á, García MD, Presa JJ, Torres B, Arnaldos MI (2015b) Micromorphology of immature stages of Sarcophaga (Liopygia) cultellata P andellé, 1896 (Diptera: Sarcophagidae), a forensically important fly. Microsc Res Tech 78(2):148–172

Varatharajan R, Sen A (2000) Role of entomology in forensic sciences. Curr Sci 78:544–546

Verma K, Paul R (2013) Assessment of post mortem interval, (PMI) from forensic entomotoxicological studies of larvae and flies. Entomol Ornithol Herpetol 2:104

Verma K, Paul R (2016) Lucilia sericata (Meigen) and Chrysomya megacephala (Fabricius) (Diptera: Calliphoridae) development rate and its implications for forensic entomology. J Foren Sci Med 2:146–150

Vincent S, Vian JM (2000) Carlotti MP partial sequencing of the cytochrome oxydase b subunit gene I: a tool for the identification of European species of blow flies for postmortem interval estimation. J Foren Sci 45:820–823

Vossbrinck CR, Coleman DC, Woolley TA (1979) Abiotic and biotic factors in litter decomposition in a semi–arid grassland. Ecology 60:265–271

Wallman JF, Leys R, Hogendoorn K (2005) Molecular systematics of Australian carrion-breeding blowflies (Diptera; Calliphoridae) based on mitochondrial DNA. InvertebrSyst 19:1–15

Wang QK, Liu XH, Lu PF, Zhang D (2014) Ultrastructure of antennal sensilla in *Hydrotaea armipes* (Fallén) (Diptera: Muscidae): new evidence for taxonomy of the genus *Hydrotaea*. Zootaxa 3790(4):577–586

Wang Y, Wang J, Wang Z (2017a) TaoI L insect succession on pig carcasses using different exposure time—a preliminary study in Guangzhou, China. J Foren and Legal Med 52:24–29

Wang Y, Wang JF, Zhang YN, Tao LY, Wang M (2017b) (2017) forensically important *Boettcherisca peregrina* (Diptera: Sarcophagidae) in China: development pattern and significance for estimating postmortem interval. J Med Entomol 54(6):1491–1497

Warren JA, Anderson GS (2013) The development of Protophormia terraenovae (R-D) at constant temperatures and its minimum temperature threshold. Foren Sci Int 233(1):374–379

Wells JD, Kurahashi H (1994) Chrysomya megacephala (Fabricius)(Diptera: Calliphoridae) development: rate, variation and the implications for forensic entomology. Med Entomol Zool 45(40):303–309

Wells JD, Sperling FA (1999) Molecular phylogeny of Chrysomya albiceps and C. rufifacies (Diptera: Calliphoridae). J Med Entomol 36(3):222–226

Wells JD, Introna F, Di Vella G, Campobasso CP et al (2001) Human and insect mitochondrial DNA analysis from maggots. J Foren Sci 46(3):685–687

Wolff M, Builes A, Zapata G, Morales G, Benecke M (2004) Detection of Parathion (o,odiethyl o-(4-Nitrophenyl) Phosphorothioate) by HPLC in insects of forensic importance in Medellin, Colombia. Anil Aggrawal's Internet J Foren Med Toxicol 5(1):6–11

Yovanovitch P (1888) Entomologie appliqué a la medicine legale||. Ollier-Henry, Paris, p 123

Zehner R, Amendt J, Krettek R (2004) STR typing of human DNA from fly larvae fed on decomposing bodies. J Foren Sci 49(2):337–340

Zou Y, Huang M, Huang R et al (2013) Effect of ketamine on the development of Lucilia sericata (Meigen) (Diptera: calliphoridae) and preliminary pathological observation of larvae. Foren Sci Int 226(1–3):273–281

Zuha RM, Razak TA, Ahmad NW, Omar B (2012) Interaction effects of temperature and food on the development of forensically important fly, Megaselia scalaris (Loew) (Diptera: Phoridae). Parasitol Res 111(5):2179–2187

Insects in Forensic Entomology

2

Abstract

Forensic entomology, an emerging subdiscipline of forensic sciences, deals with critical analysis of carrion associated fauna to ascertain the PMI and related issues of medicolegal relevance. Though forensic entomological research has gradually gained pace in the past decade in India, yet, it's still to make a mark in a nation as vast and diverse like India, with different geographical territories and with variable climatic conditions and seasonal fluctuations. In this chapter we highlight forensic entomological research with focus of study from India, emphasizing on the three predominant families of Necrophagous Diptera, namely, Calliphoridae, Sarcophagidae, and Muscidae.

Keywords

Necrophagy · Forensic entomology · Blow flies · Flesh flies · Muscids

2.1 Introduction

Entomology involves the study of insects, belonging to invertebrate phylum Arthropoda, class Insecta wherein the body encompasses an exoskeleton, with six jointed legs, three body segments, head with one pair of antenna, and one pair of compound eyes.

Forensic entomology encompasses the application of study of arthropods and insects in study and analysis of legal matters. Insects that find application in forensic science includes two major groups under the broad heading of flies (Diptera) and beetles (Coleoptera), while their Arthropod relatives include non-insects like mites, spiders, and ticks. The first ever study on forensic entomology dates back to 1986 with the publication of the first manual (Smith 1986). Since then, interest in this domain and relentless work by scientist and technologist have further enriched this

S. Ghosh, D. Banerjee, *Biology of Forensically Important Invertebrates*,
https://doi.org/10.1007/978-981-99-5026-3_2

domain that has contributed to increased robustness of the interpretation of insect evidence.

In the domain of forensic sciences, forensic entomologists enable detection of the postmortem interval (PMI) calculated based on known period of insect activity, or time since colonization post-death on a dead body.

They also find importance in establishing the geographical location of death in cases of body transport and help in detection of victim and suspect and their relation to the site of death. They also play a role in aiding the pathologist to identify sites of trauma on the dead body. Entomology finds applications in detection of toxicology samples, and the gut contents of such forensic flies find applications in human DNA viable for forensic testing.

A forensic entomologist detects the time at which the insect had colonized the dead body for the first time after it appeared on the corpse first, thus inferring at the minimum PMI (PMImin). Such data is often used together with data from chemical, histological, and bacteriological studies to establish time elapsed since death (Easton and Smith 1970).

Conventional methods of detection based on changes postmortem in a cadaver measured by pathologists thereby used in estimation of PMI like rigor mortis, algor mortis, liver mortis suffer from lack of reliability due to lack of exact and accurate accounting and methods used are largely subjective rather than being objective (Henssge et al. 1988).

Since most postmortem changes used to estimate PMI occur within the first 72 h of death, the estimate by the medical examiner lies restricted to the first 3 days, and usage of a single accurate method remains largely debatable (Swift 2010).

But events of egg laying of the first blow flies on a corpse start a biological clock; thus, this objective method finds importance in study of colonization and development of insects. Estimation of time of death can range from a few weeks in summer and a few months in cooler seasons.

Even in conditions where the body has been decayed and discovered late, the study and identification of the succeeding colonizers and insect fauna including flies and beetles that accompany with changes of the body with decomposition still provide a useful information but no exact time of death (Greenberg and Kunich 2002).

The study of insect succession on corpse records back in in the late 1800s (Megnin 1894) and is still being studied with great interest as important topic of research (Matuszewski et al. 2010). However, local and seasonal changes remain the major limitations that lead to difficulty in interpretation of data.

2.2 Major Areas of Application of Forensic Entomology

Forensic entomology with its vast applications in deciphering legal matters (Hall 2001, 2012) includes three major areas of application (Lord and Stevenson 1986) including:

Table 2.1 Major areas of application of forensic science

Types	Applications
Urban entomology	Insects and human environments
Food/stored products entomology	Insects and their presence in food products
Medicolegal/criminal entomology	Criminal proceedings in detection of crime or death

(i) Urban entomology, with reference to insects and human environments
(ii) Food/stored product entomology relating to insects occurring in food products
(iii) Medicolegal/criminal entomology finding importance in criminal proceedings in detection of crime or death (Hall 1990; Hall et al. 2008, 2012) (Table 2.1).

Application of entomology in the medicolegal/criminal field has been developed most widely in the past few decades that find applications in different circumstances like study and detection of

(i) Isolated domestic/local incidents/deaths
(ii) Large-scale crimes/murder
(iii) Different investigation purposes like neglect of elderly (Benecke 2004)
(iv) Events of child abuse (Benecke and Lessig 2001)
(v) Detection of illegal events of poaching of wildlife (Anderson 1999; Samuel 1988)
(vi) Detection of gun-shot events (Roeterdink et al. 2004)
(vii) Use of drugs or entomotoxicology (de Carvalho 2010)
(viii) Movement of vehicles, based on insects observed and identified on windscreen glass of vehicle
(ix) Insects observed and identified on windscreen glass of vehicle used in transportation and relocation of human corpse/remains or dead bodies of animals (Smith 1986) and thereby estimation of minimum time that has elapsed since death (Hart et al. 2008).

Extensive studies by Amendt et al. (2010), Byrd and Castner (2001, 2009), Haskell and Williams (2008), Erzinclioglu (1989, 2000), Gennard (2007), Goff (2000), and Greenberg and Kunich (2002) in this domain of biology have highlighted the association of insects with human dead bodies.

2.3 Forensically Important Insects: Ecology

These insects have been studied in the ecological groups under the following headings:

(a) **Necrophagous** species of insects that feed on the body of dead or decaying animal flesh reveal information and evidence as they possess a direct and primary association with the body. Scavenger vertebrates, including vultures

Table 2.2 Role of detritivores (Edwards et al. 1970; Kitchell et al. 1979)

1. Accelerate microbial decomposition
2. Nutrient cycling
3. Maintenance of homeostasis in the ecosystem by shredding, consuming, and transforming organic detritus

and jackals, contribute to the detritivore community by devouring remains of dead animals. The necrophagous arthropod community finds importance from ecological perspectives (Amendt et al. 2004; Anderson and VanLaerhoven 1996, Moore et al. 2004; Seastedt and Crossley 1984) (Table 2.2).
Carrion represents transient microhabitats that vary according to the abiotic environmental factors as well as the species complex that dominates the carrion from one patch to another. The role each species of carrion flies plays within a specific climatic zone defines its niche and also its relationship with other coexisting species. Thus, accumulating evidences from previous studies highly confer rapid ecological successions within a carrion microhabitat because of the mass degradation of carrion by the arthropod micro-fauna that colonize it (Galante and Marcos-Garcia 2008).

The diversity of arthropods that dominates carrion forms the carrion community which is far more complex than being considering the role of only one species or population. These complex carrion dwellers include several species of the order Diptera, Coleoptera, Arachnida, and Hymenoptera (Amendt et al. 2000, 2004; Gruner et al. 2007).

One more important characteristic feature of carrion succession outlines the dominance and distribution of diverse insect complexes at different stages of cadaveric decomposition which in turn appears as waves of succession (Anderson 2010; Byrd and Castner 2001). Specifically, the mouthparts and antennal sensilla of carrion flies contain a wide array of highly developed olfactory receptors for detecting pungent odor of various ammonia and sulfur-rich chemicals released from the rotten flesh matter which in turn also provides bed for massive bacterial growth along with the release of chemical by-products indicating the amount of decomposition (Joseph et al. 2011; Mondor et al. 2012; Bhattacherjee et al. 2021a, b, c; Hore et al. 2017a, 2018). The chemicals released during different stages of decomposition also regulate a wide range of behavioral facets in cadaveric insect fauna (LeBlanc and Logan 2010; Joseph et al. 2011; Bhattacherjee et al. 2021a, b, c). Therefore, being attracted by the peculiar odor of volatile chemicals released during decomposition, vast insect communities colonize the carcass as well as largely determine the precise stage of decomposition. Evidently, the sulfur-rich chemicals attract carrion flies, while volatile ammonia and its derivatives stimulate oviposition efficacy during early stages of cadaveric decomposition (Ashworth and Wall 1994; Joseph et al. 2011; Bhattacherjee et al. 2021a, b, c).

The foraging behavior of necrophagous dipterans allows them to colonize the carrion microhabitat during the onset of decomposition process, owing to their well-equipped sensillary system. Thus, they are found to be the first-line colonizers of corpses within a few minutes (Anderson and Val 2001; Dadour et al. 2001; Greenberg and Kunich 2002; Byrd and Castner 2009; Bhattacherjee et al. 2021a, b, c; Hore et al. 2017a, b) and sometimes even a few seconds (De Jong 1995) after the death of an animal. The decomposition is one of the key events that maintain a functional balance within any ecosystem. The larvae of the adult Diptera which oviposit on rotten flesh matters are regarded as the most rapacious feeders and are the masters of recruiting the bacterial fauna that promotes liquefaction of the cadaveric tissues (Galante and Marcos-Garcia 2008). However, the adult necrophagous flies are found to be fluid feeders as they suck the fluids/sap directly from the cadaveric tissues.

Though the adult dipterans feed on the fluids of the corpse, the larvae are the true drivers of the decomposition process, secreting enzymes directly into the carrion, thus facilitating carrion decomposition.

Decomposers help in nutrient cycling and maintaining ecological balance. Owing to the vital role played by the carrion feeder flies during cadaveric decomposition, they are considered as potent nutrient cyclers that contribute in maintaining proper homeostasis within any terrestrial ecosystem (Allee et al.1949; Galante and Marcos-Garcia 2008; Kuusela and Hanski 1982; Hore et al. 2017a).

Depending on the type of ecosystem, producers, consumers, and decomposers are found to be the major organisms that comprise any ecosystem considering their biomass and energy consumption at corresponding trophic levels inhabited by them. In addition, by their intra-group and inter-group interactions, the role of these crucial players is measured (Galante and Marcos-Garcia 2008).

The availability of nutrients at different trophic levels is highly facilitated by the action of decomposers as they assist in cycling of nutrients to maintain the functional interaction with producers and consumers (Galante and Marcos-Garcia 2008).

From recruiting potent bacterial communities to humus formation in the soil, decomposers make the nutrients available to the plant communities. Additionally, it also has remarkable roles in carbon and other nutrient cycling with effects at the ecosystem level as the resources that enter a detrital food web are largely controlled by these small-sized species (Yang and Gratton 2014).

Evidently, these necrophagous flies not only are important members of detritivore community that involves the consumption of carrion by larval forms but also they make beds for the proliferation of huge number of microorganisms which in turn facilitates the process of nutrient recycling by making the nutrients available to the plant species (Galante and Marcos-Garcia 2008; Hanski 1987).

(b) **Necrophagous Flies as Potential Bioindicators**

There are several necrophagous species of dipteran flies which act as potent bioindicators to measure the amount of environmental changes and its consequences including forest restoration (Majer 1987).

Dipterans owing to their high amount of adaptability are found in diverse types of habitats and interact with several abiotic and biotic environmental factors cautiously culminating in vast ecological diversity of both the adult and larval forms (D'Almeida and Lopes 1983). The role they play in diverse habitats (wild as well as anthropogenic environments) at different trophic levels determines their realized niche (Majer 1987; D'Almeida and Lopes 1983).

However, some dipteran species discretely are often circumscribed at narrower environmental conditions (Odat et al. 2015; Pereira de Sousa et al. 2014); thus, when environmental parameters become a bit unsuitable, dipterans act as communities to invade specific habitats.

Not only the necrophagous species complex but also several dipteran communities as a whole serve as bio-sensitive indicators for detecting even elusive fluctuations in the environmental parameters of a particular zone. There are several families of forensic flies such as Calliphoridae and Sarcophagidae which largely vary in their diversity in response to anthropogenic disturbances. Evidently, both families are found in zones where man-made disturbances and industrialization are intermediate, while the diversity remarkably declines if there are heavy disturbances or no disturbances at all (Dufek et al. 2016).

To conclude their role as potent environmental indicators, several renowned research groups have performed precise investigations on their behavior as well as ecology.

(c) **Dipteran Families of Forensic Importance**

Dipterans dominate carrion conveying exceptional role in nutrient recycling and decomposition. The waves of appearance on cadaveric remains commence with Calliphorid species followed by Sarcophagidae and Muscidae within a few minutes or a few hours in most cases. By modifying the conditions of carcasses physically and chemically, the foremost species groups allow the settlement of species that arrived at later stages of the decomposition (Ricklefs 1996; Ames and Turner 2003). Calliphoridae species are necrophagous and not only breed in carrion but also on excrement or other types of organic matter and may act as vectors of several animal and human diseases. Sarcophagid species of forensic importance are mostly known to colonize human remains, while Calliphorid species are equally associated with garbage and animal excrements. In India, however, most of the forensic entomological works have been performed taking Calliphorid flies into account similar to South America probably due to the climatological similarity of both regions (Wells and LaMotte 2001).

Necrophagous dipteran flies are known to visit cadavers and carcasses soon after the death of the animal. The succession of carrion flies occurs in waves with variation of species diversity during each successive stages of decomposition of the cadaver. Furthermore, the necrophagous species complex that dominates a particular type of carcass also varies with the environmental parameters as well as with the stage of decomposition of the carrion tissue mass. Among the several species of necrophagous flies that colonize a carcass, the blow fly of the family Calliphoridae are generally the first group to visit carrion. Markedly, they appear within minutes after the animal dies followed by other families of necrophagous species such as Sarcophagidae, Muscidae, and subsequently by other less dominant families of necrophagous flies (Amendt et al. 2004).

Furthermore, investigating the colonization as well as succession patterns of dipteran flies on a specific type of carcass provides hints about whether the body has been displaced from an urban to a rural habitat or vice versa (Erzinclioglu 1989; Catts and Haskell 1990; Bhattacherjee et al. 2021a, b; Hore et al. 2017b).

Among the several different families of necrophagous flies, Calliphoridae, Sarcophagidae, and Muscidae are found to be the most important forensic indicators on a global scale as they contribute profoundly in detecting postmortem interval (PMI) since death. However, there are also some other families of flies that visit carcass and rarely colonize it. These groups of flies belong to families Psychodidae, Sepsidae, Neriidae, Stratiomyidae, Piophilidae, Drosophilidae, Phoridae, Fanniidae, Ephydridae, Chloropidae, Milichiidae, Sphaeroceridae, Micropezidae, and Cecidomyiidae (Amendt et al. 2004; Cavallari et al. 2015; Hore and Banerjee 2017; Ndueze et al. 2013; Pereira de Sousa et al. 2014, 2015, 2016; Rochefort et al. 2015).

Considering the Indian scenario, the most prevalent forensic species of dipterans include Calliphoridae, Sarcophagidae, and Muscidae as these families of flies are mostly recorded from cadaveric remains throughout different parts of India. However, there are remarkable differences in their species composition according to the climatic variables. The minor group of carrion visitor flies from India includes dipteran families like Phoridae, Fannidae, Piophilidae, lidiidae, and Anthomyiidae (Bharti and Singh 2003; Majumdar et al. 2007; Sinha 2009; Chakraborty et al. 2015a, b, 2016a, b; Singh et al. 2016; Hore and Banerjee 2017).

(d) **Forensic Entomology and Necrophagous Flies**

Necrophagous flies have extraordinary implication in forensic entomology as they help in the estimation of postmortem interval (PMI) since the death has occurred. During the first phase of a few hours, chemical and physical methods serve as the prime method of PMI detection. However, when the animals or human dead bodies are found after longer periods of time have passed, the use of the physical and chemical methods of PMI detection doesn't work. Thus, the mode of PMI detection that the forensic investigators and criminologists practice involves the biological method of PMI detection, and if the biological method deals with the association of forensic insects, then the process is known as forensic entomology.

Pathologists can precisely reckon the time since the death during the first few hours, but when the duration of retrieving a dead body becomes longer than 72 h, conditions, for instance, rigor mortis/livor mortis and even body temperature, become unsuitable strategy for the accurate detection of the time of death. Evidently, the species complex that belongs to a particular habitat contributes to the accurate estimation of minimum postmortem interval (PMI) (Amendt et al. 2004; Kashyap and Pillai 1989).

Apart from the role of forensic flies in estimating PMI, the wave of insect succession pattern associated with a particular carcass needs analysis to determine how they add notable significance in carrion ecology as well as the influence of necrophagous Diptera fauna has important implications in estimating several environmental parameters in connection with the decomposition of cadavers (Byrd and Castner 2001). Furthermore, the fact is that several species of insects require precise environmental parameters for colonization of a particular carcass. In addition, the successional waves of necrophagous flies provide indications about the relocation of the decomposed remains from a rural to an urban habitat and vice versa (Erzinclioglu 1989; Catts and Haskell 1990; Hore et al. 2017b). Several retrospect data regarding the succession patterns of forensic flies on cadaveric remains shed light on the specific role of these forensically important families of flies as forensic indicators by facilitating PMI estimation (Azmi and Lim 2013; Bhattacherjee et al. 2021a, b; Carvalho et al. 2004; Heo et al. 2008; Hore 2021).

According to previous research reports, the theory of fully matured maggots that voraciously feed on cadaveric remains can also provide clear indications of how to determine the PMI. For the later purpose, developmental analysis of oldest maggots seems to be critical (Amendt et al. 2004).

By the precise method of forensic entomological analysis, human death cases can be solved when physical and chemical methods fail to accomplish proper investigation due to the lack of circumstantial evidences. However, forensic entomology also aids in wildlife crime control; thus, these small creatures are the operators that accelerate the process of crime investigation (Anderson 1999).

Obviously with microhabitat specificity, carrion and its interaction with different environmental parameters differ broadly. A clear picture of the base of any forensic study needs proper analysis of the climatic variables and several other factors such as habitat patches, area, and seasons (Matuszewski et al. 2010). The latter factors are markedly related with the decomposition rate of the carrion. One more important factor is anthropogenic activities in a particular habitat that essentially shape the species complex. Thus, the species composition of forensic dipterofauna in each region differs substantially. Concerning forensic entomological investigation in each region, the geographical and ecological viewpoint stands out to set the foundation of estimating decomposition rate of remains, and the details of the species composition of insect fauna from each region need to be deduced with utmost promptitude for assigning variances in decomposition rates.

(e) **Predators and parasites** of necrophagous species include insects that feed on, or are parasites on, necrophagous species. They do not feed directly on the decomposing remains or its fluids but act as second in forensic insect in ecological role. Species from the Coleoptera families including Silphidae commonly called as carrion beetles and Staphylinidae commonly called as rove beetles, and parasitic wasps, under order Hymenoptera under family Braconidae are examples. Some blow fly species may initiate their larval development in the necrophagous role, feeding directly on remains, but during later larval stages may become predaceous and exhibit schizophagous nature.

(f) **Omnivorous species** include those species that feed on the dead decaying body and other inhabitant insects associated with the carrion-like necrophagous species. They in large numbers can slow down the rate of removal of carcass materials by depleting necrophagous larvae. Anta, wasps, and some species of carrion beetles fall under this category.

(g) **Adventive species** use the dead decaying body as their extended environment (Smith 1986) using it as their own natural habitat. This is observed in Collembola, springtails, spiders, and centipedes. Acari, and families including Acaridae, Lardoglyphidae, and Winterschmidtiidae, thrive by feeding on developing molds and fungi on the dead body. They may/may not play a significant role in the decomposition process. They may visit remains from time to time, or use for hiding.

2.4 Blow Flies

Among all groups, blow flies (Diptera: Calliphoridae) (Fig. 2.1a–e) find importance in forensic entomology as they are they are the first group to colonize a dead body in largest numbers, and therefore they provide the most accurate information as regards the minimum time elapsed since death (Greenberg 1991). Blow fly adults and their

Fig. 2.1 (**a–e**) Some Indian blow fly species of forensic relevance. (**a**) *Chrysomya megacephala*, (**b**) *C. rufifacies*, (**c**) *C. nigripes*, (**d**) *Hemipyrellia ligurriens*, (**e**) *Lucilia cuprina* (Reproduced with permission, Hore, 2021, and Director, ZSI, India)

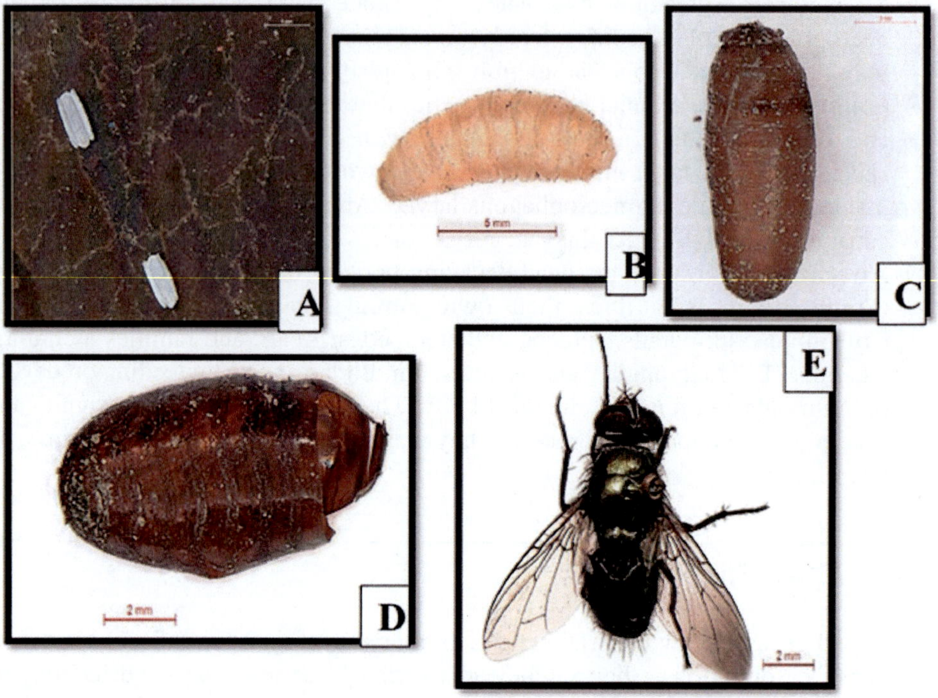

Fig. 2.2 (**a**–**e**) Calliphorid fly, *Hemipyrellia ligurriens*. (**a**) Eggs (shown with black arrows). (**b**) Third instar larva. (**c**) Pupa (dorsal view). (**d**) Empty puparium (ventral view). (**e**) Adult. (Reproduced with permission from- Hore, 2021, and Director, ZSI, India)

different developmental stages including egg, larva/maggot, and pupa (Fig. 2.2a–e) find importance in predicting the minimum time elapsed after death.

(i) The larva is the main feeding stage and it includes the three instars (first, second, and third) separated by a molt (shedding) of the cuticle (skin) allowing for growth.
(ii) Post-completion of feeding, the larvae move out of the body and burrow into the nearby environment like soil, or under furniture and carpets/curtain, etc.
(iii) Post-metamorphosis, from larva the adult fly develops.
(iv) Pupal stage development takes place in the puparium, the hardened and darkened cuticle of the third instar larva.

Blow flies including a large family of flies, secondary screwworm fly (Cochliomyia macellaria), Arctic blow fly (*Protophormia terraenovae*), green bottle fly (*Lucilia* sp.), blue bottle fly (*Calliphora vomitoria*), and bronze bottle fly (*Lucilia cuprina*) find importance in forensic entomology.

Analysis of immature insects is crucial in forensic entomology, as these immature forms of carrion entomofauna are generally encountered upon and collected during

Fig. 2.3 Third instar larva of the Calliphorid fly, *Chrysomya rufifacies*. (Reproduced with permission from Hore, 2021, and Director, ZSI, India)

forensic investigations, and so, accurate identification of the involved species is the foremost step in calculating the postmortem interval (PMI) (Thyssen 2009).

The immature stage comprises the egg, maggot or larva with its average of three or four development instars, and the pupa (Thyssen 2009).

(i) **Adult flies/insects** (Fig. 2.2e) lay large eggs in large numbers that hatch rapidly. The larvae/maggots develop into multiple instars larva and then form puparia leading to emergence of adult. The life cycle and timing of each of these growth stages are well studied, and this information finds importance in detecting the time since death. Climate parameters like temperature, season, humidity, etc. are noted also (Ubelaker 2013).

(ii) **Eggs** (Fig. 2.2a) form the first stage of development in which a series of changes takes place before the larvae hatches, as the embryos contain cells and developmental programs for larval structures (Thyssen 2009).
During death scene investigations, the forensic investigator is likely to encounter insect eggs while collecting entomological evidence (Byrd and Castner 2001). Dipteran eggs are commonly observed as clusters or masses of fly eggs that accumulate around the natural body orifices such as the eyes, nose, mouth, genital/anal area, and around wound sites, if any (Byrd and Castner 2001). Collectively and singularly, in appearance, the dipteran eggs are characteristically, creamy-white to yellowish in color, and dry to touch and will easily "flake" away from the substrate on which they were deposited by the female (Byrd and Castner 2001). The short duration of this stage makes it a particularly valuable life cycle stage for the forensic entomologist to recover and analyze, crucial in rendering precise PMI estimation (Byrd and Castner 2001).

(iii) **Larvae** (Figs. 2.2b and 2.3): The larvae of these flies are widespread and inhabit terrestrial, aquatic bottom, water column of lakes, streams, and ponds and soil like sand, dead wood, dung, carrion, and decaying vegetation and inhabit social insect colonies. Alive larvae of herbivorous tipulids and many aphidophagous syrphids have been reported inhabiting on vegetation. Larva of

Diptera reveals diverse feeding habits including predators, scavengers, herbivores, parasitoids, or even true parasites.

(a) **Scavengers**

Diptera larvae, including Brachycera, reveal scavenger-like existence feeding on dead decaying organic substances, deriving their nourishment from decayed materials by bacteria and other microbes. These bacterial digestion products and suspended particles get concentrated through a sieve-like pharyngeal filter in the larva. Calliphoridae larvae are most common terrestrial scavengers found on fresh dead animal carcasses. Calliphorid and Muscid larvae are the first organisms in early-stage decay, followed by drosophilids, fanniids, phorids, piophilids, sepsids, sphaerocerids, and others.

Although Calliphorids are not reported from buried dead bodies, muscid genus *Muscina*, phorids, and sphaerocerids are reported to be predominate.

Similarly, faunal composition of carrion immersed in water is different from those studied in dry land. Decaying vegetable matter has been reported to be rich with fauna of muscids, sphaerocerids, sciarids, and others, while decaying seashore vegetation reveals growth of coelopids, sphaerocerids, and anthomyiids. Drosophilids, stratiomyids, and some phorids are reported from rotting fruits. In aquatic environments, many larval Diptera feed on small organic particles in the water and substrate. Larvae of Culicidae and Simuliidae are known to use their brush-like labral fans to filter particles from the water.

(b) **Herbivores**

Diptera feed on living algae and fungi which include Agromyzidae, Anthomyiidae, Bibionidae, Cecidomyiidae, Chloropidae, Phoridae, Psilidae, Sciaridae, Tephritidae, and Tipulidae and are reported to be pests to human agriculture. Fly larvae attack plant parts including fruits, flowers, stems, leaves, and roots.

(c) **Predators**

Predators including non-cyclorrhaphan Brachycera reveal active larval stages attacking other invertebrates as food. The tabanid larva have been reported to kill frogs. These predatory dipteran larvae like syrphids killing aphids and muscid larvae killing other muscid larvae like houseflies find importance in controlling insect pests.

(d) **Parasitoids**

Parasitoids are known to develop on and kill a single host which include 22 families of Diptera reviewed by Feener and Brown in 1997 (Feener Jr. and Brown 1997). All species of the Tachinidae are parasitoids. All dipteran parasitoids are known to attack other invertebrates, like insects, but unusual hosts include terrestrial flatworms, mollusks, earthworms, millipedes, spiders, and scorpions.

(e) **Parasites**

Parasites feed on a single host but do not kill it. Larvae of bot flies survive in the dermal layers of human, nasal cavities of various mammals, or stomachs and alimentary tracts of horses, elephants, zebras, and rhinos. Calliphorids, chloropids, piophilids, and muscids larva act as parasites of vertebrates. Phoridae includes true parasites of invertebrates.

(f) **No Free-Living Larvae**

The larvae families including Hippoboscoidea, Glossinidae, Hippoboscidae, Nycteribiidae, and Streblidae survive in the female abdomen and nourished by secretions of the accessory glands, and the mature third-instar larvae are deposited by females (Brown 2001).

(iv) **Pupa** (Fig. 2.2c, d): The puparium of the dipterans belonging to infraorder Muscomorpha is generally characterized as barrel-shaped, heavily sclerotized (which is formed by hardening of the third instar larval cuticle), being morphologically comparable to the previous larval instar, yet smaller in length due to a retraction in the body segments (Thyssen 2009).

2.5 Flesh Flies

Flesh flies, members of family Sarcophagidae, are a large family of flies, worldwide in distribution, with majority of the species prevailing either in tropical or warm temperate regions (Byrd and Castner 2001). Morphologically, the adults generally have gray and black longitudinal stripes on the thorax and a characteristic (checkerboard) pattern on the abdomen (Byrd and Castner 2001). Etymologically, the family's Latin name denotes "flesh eating" and seemingly refers to the larvae or maggots that usually thrive on animal material (Byrd and Castner 2001). Besides carrion, they are also reported to feed on excrement or exposed meats (Byrd and Castner 2001).

Flesh flies have been reported to visit carcasses under most conditions, including sun, shade, dry, wet, indoors, as well as outdoors (Byrd and Castner 2001). They can be found associated with carcasses throughout, both during the early and advanced stages of decomposition (Byrd and Castner 2001; Hore 2021; Ren et al. 2018). Female flies in this family do not lay eggs, and, thus, fly egg masses associated with human cadavers are not found in sarcophagids (Byrd and Castner 2001); instead, they are viviparous, that is, deposit living first instar larvae on decomposing remains (Byrd and Castner 2001). Thus, the time span crucial for egg development ought to be ruled out while calculating a postmortem interval (PMI) based on flesh fly evidence (Byrd and Castner 2001).

Relevant from the forensic entomological point of view, in comparison to other fly species, sarcophagids possess distinct features facilitating the estimation of PMImin (Ren et al. 2018). First of all, the reproductive nature of ovoviviparity (or ovolarviparity) in majority of flesh flies enables them to deposit maggots directly on a carcass, instead of eggs (Byrd and Castner 2001; Singh and Bharti 2008; Szpila et al. 2015). Secondly, they are readily visible than others owing to their larger size (Cai 2015; Pape 1996; Tomberlin and Benbow 2015). Lastly, sarcophagid flies are reported to be more active in various stages of decomposition of carcasses, thus are key players aiding forensic investigations (Anderson and VanLaerhoven 1996; Byrd and Castner 2001; Povolny and Verves 1997; Silahuddin et al. 2015).

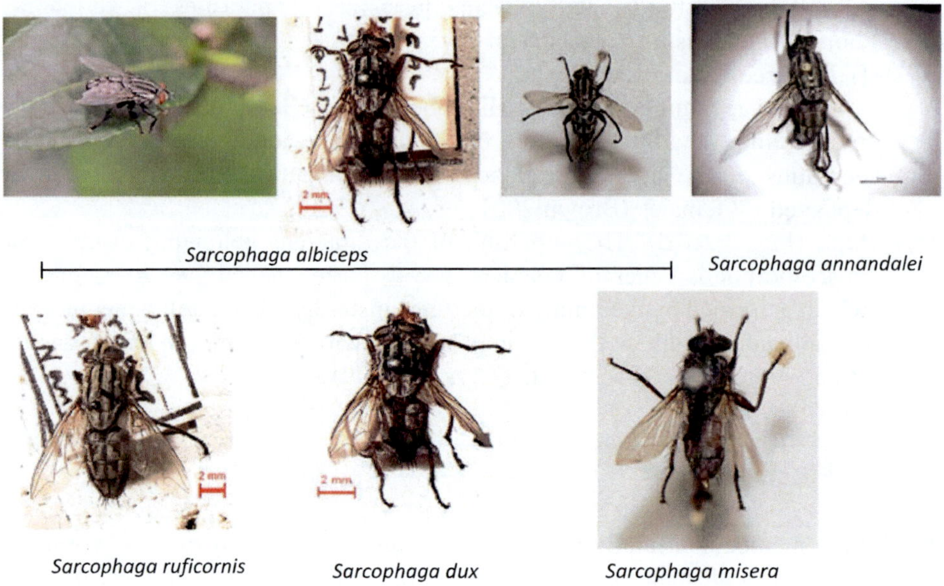

Fig. 2.4 Indian varieties of *Sarcophaga* sp. (Reproduced with permission from Director, ZSI, India)

Flesh flies belonging to genus *Sarcophaga* are primarily of forensic relevance worldwide and arrive on cadavers simultaneously, or a bit later than the blow flies (Byrd and Castner 2001). Common sarcophagid species of forensic relevance in India are *Sarcophaga (Liopygia) ruficornis, Sarcophaga (Liosarcophaga) dux,* and *Sarcophaga* (*Parasarcophaga*) *albiceps* (Nandi 2002) (Fig. 2.4).

2.6 Muscid Flies

A large family of true flies, with worldwide distribution pattern, family Muscidae encompasses members such as house flies, stable flies, and horn flies, many of which are essentially synanthropic in nature (Byrd and Castner 2001). Owing to their synanthropic nature, some species of muscids are apparently involved in medical and forensic cases (Byrd and Castner 2001; Smith 1986) (Figs. 2.5 and 2.6).

Muscids tend to colonize carcasses after the flesh flies and blow flies (Byrd and Castner 2001). They are reported to often oviposit in natural body orifices, at wound sites, or in fluid-soaked clothing (Byrd and Castner 2001), as well as beneath the body into soil soaked by decomposition liquid (Šuláková 2014). Larvae thrive directly on carrion, though predacious behavior has been displayed by some species as they mature (Byrd and Castner 2001). In such instances, the faunal composition on a set of remains may be affected as muscid larvae tend to devour the eggs and larvae of other carrion-feeding flies (Byrd and Castner 2001).

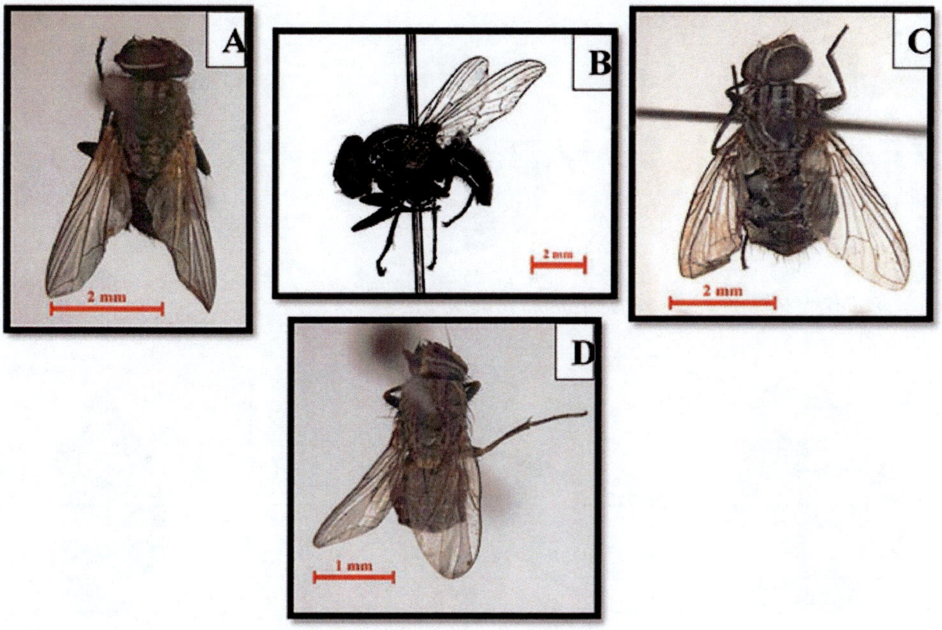

Fig. 2.5 (**a–d**) Some Indian Muscid species of forensic significance. (**a**) *Musca domestica*, (**b**) *Ophyra capensis*, (**c**) *Synthesiomyia nudiseta*, (**d**) *Atherigona orientalis*. (Reproduced with permission from Hore, 2021 and Director, ZSI, India)

Common members of this family of worldwide forensic relevance as well as from the Indian perspective are *Musca domestica* Linnaeus, 1758, *Musca (Byomya) sorbens* (Wiedemann, 1830), *Atherigona (Acritochaeta) orientalis* (Schiner, 1868), *Synthesiomyia nudiseta* (van der Wulp, 1883), *Ophyra capensis* (Wiedemann 1818), and *Hydrotaea chalcogaster* (Wiedemann 1824) (Byrd and Castner 2001; Grzywacz et al. 2017; Hore 2021; Joseph and Parui 1980; Sinha 2009).

2.7 Beetles

Together with blow flies, other insects like hide beetle (Dermestes) that colonize mostly dried bodies find importance in estimating a minimum time elapsed post-death with known rates of development (Kumara et al. 2009; Midgley et al. 2010).

Among different arthropods involved in human decomposition, flies (Diptera) and beetles (Coleoptera) are attracted to moist tissue and thus arrive early post-death. The fly larvae act on the soft tissue and cause reduction. In Arthropod succession, beetles appear later, as they are attracted by more desiccated tissue. Most other arthropods appear to consume feeding insects. A forensic entomologist can identify and collect adults, eggs, and larvae and use that information to assess time passed since death.

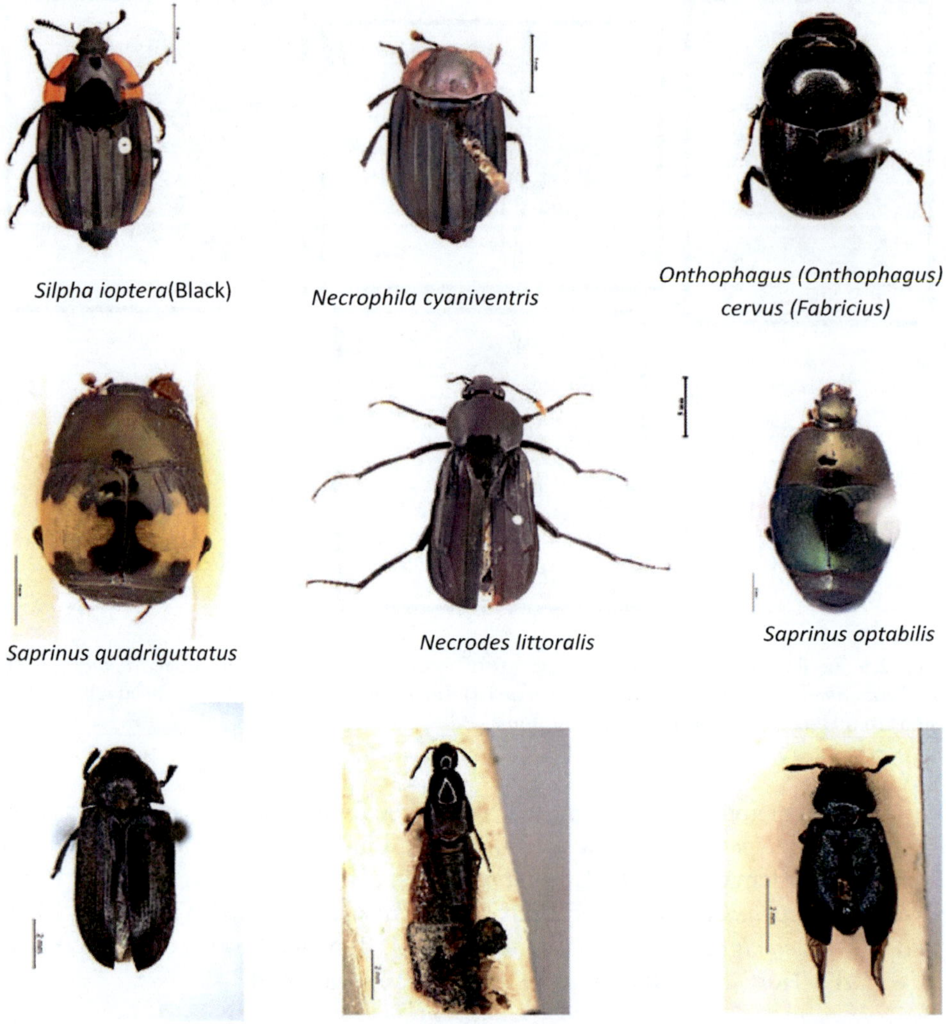

Silpha ioptera(Black) *Necrophila cyaniventris* *Onthophagus (Onthophagus) cervus (Fabricius)*

Saprinus quadriguttatus *Necrodes littoralis* *Saprinus optabilis*

Dermestes maculatus *Philonthus_longicornis* *Necrobia rufipes*

Fig. 2.6 Different members of Coleoptera beetles of forensic importance. (Reproduced with permission from Dr. Devanshu Gupta, Scientist-D, ZSI and Director ZSI)

2.8 Discussion

In medicolegal scenarios, the time of death is the key question asked by a forensic entomologist especially in death of human or animal life and in cases of suspected crimes/untimely/unnatural death/illegal poaching (Wells and Lamotte 2001), and it is important to estimate the time elapsed past death, or postmortem interval (PMI). Keen interpretation of carrion succession patterns of necrophagous entomofauna not only has wide applications in ecological and environmental sciences but also in PMI estimation in medicolegal entomology when carrion arthropods are collected and resolved as evidences (Byrd and Castner 2001). The carrion arthropod species composition can differ among habitats, regions, days, and seasons (Matuszewski et al. 2010). Hence, for attaining accurate results in forensic entomological studies, it is crucial that the study of insect fauna in each region is conducted for interpreting differences in decomposition of carrion in relation to biogeography and ecology of necrophagous insect communities. The present review is aimed at providing a basic concept regarding the fundamentals of forensic entomology and carrion entomofauna associated with it, particularly emphasizing the Indian perspective.

References

Allee WC, Emerson AE, Park O, Park T, Schmidt KP (1949) Principles of animal ecology, 1st edn. WB Saunders Co. Ltd., Philadelphia, p 837

Amendt J, Krettek R, Niess C, Zehner R, Bratzke H (2000) Forensic entomology in Germany. Forensic Sci Int 113(1–3):309–314

Amendt J, Krettek R, Zehner R (2004) Forensic entomology. Naturwissenschaften 91:51–65

Amendt J, Goff ML, Campobasso CP, Grassberger M (eds) (2010) Current concepts in forensic entomology. Springer, London

Ames C, Turner B (2003) Low temperature episodes in development of blowflies: implications for postmortem interval estimation. Med Vet Entomol 17(2):178–186

Anderson GS (1999) Wildlife forensic entomology: determining time of death in two illegally killed black bear cubs. J Forensic Sci 44(4):856–859

Anderson GS (2010) Factors that influence insect succession on carrion. In: Byrd JH, Castner JL (eds) Forensic entomology: the utility of arthropods in legal investigations, 2nd edn. Boca Raton, FL, CRC Press, pp 201–250

Anderson G, Val C (2001) Insects associated with the body: their use and analyses. In: Advances in forensic taphonomy. Methods, theory and archeological perspectives. CRC Press. https://doi.org/10.1201/9781420058352-12

Anderson G, VanLaerhoven SL (1996) Initial studies on insect succession on carrion in Southwestern British Columbia. J Forensic Sci 41:617–625

Ashworth JR, Wall R (1994) Responses of the sheep blowflies Lucilia sericata and L. cuprina to odour and the development of semiochemical baits. Med Vet Entomol 8(4):303–309

Azmi WA, Lim SP (2013) Comparative study of dipteran species diversity and their succession on rabbit carrion in two different mangrove areas of peninsular Malaysia. J Insects 2013.: Article ID 398159:9. https://doi.org/10.1155/2013/39815

Benecke M (2004) Arthropods and corpses. In: Tsokos M (ed) Forensic pathology reviews. Humana Press Inc., Totowa, NJ, pp 207–240

Benecke M, Lessig R (2001) Child neglect and forensic entomology. Foren Sci Int 120:155–159

Bharti M, Singh D (2003) Insect faunal succession on decaying rabbit carcasses in Punjab, India. J Foren Sci 48(5):1133–1143

Bhattacherjee R, Ghosh S, Banerjee D (2021a) The chemosensory system of forensic flies controls their biophysiological behavior. Insect Environ 24(3):439–441

Bhattacherjee R, Ghosh S, Banerjee D (2021b) Scanning electron microscopic study of the antennal sensilla of female *Sarcophaga orientaloides* (Senior White, 1924), a flesh fly of forensic importance and its association with a filamentous fungus. J Ent Res 45:1080–1089

Bhattacherjee R, Banerjee D, Ghosh S (2021c) Chronology of appearance pattern and colonization of nutrient recycling forensic flies on house rat carcasses (rattus rattus) in urban and semiurban habitats of southern part of West Bengal. Int J Entomol Res 6(2):148–156

Brown BV (2001) Flies, gnats, and mosquitoes. In: Levin SA (ed) Encyclopedia of biodiversity, 2nd edn. Academic Press, pp 488–496. ISBN 9780123847201
Byrd JH, Castner JL (2001) Insects of forensic importance. In: Forensic entomology: the utility of arthropods in legal investigations, 2nd edn. CRC Press, pp 39–126
Byrd JH, Castner JL (2009) Forensic entomology: the utility of arthropods in legal investigations, 2nd edn. CRC Press, Boca Raton, FL
Cai JF (2015) Forensic entomology. People's Medical Publishing House, Beijing
Carvalho LML, Thyssen PJ, Goff ML, Linhares AX (2004) Observations on the succession patterns of necrophagous insects on a pig carcass in an urban area of Southeastern Brazil. Forensic Med Toxicol 5(1):33–39
Catts EP, Haskell NH (1990) Entomology and death: a procedural guide. Forensic entomology specialties, Clemson, SC
Cavallari ML, Baltazar FN, Nihei SS, Muñoz DR, Tolezano JE (2015) Dipterofauna associated with SusscrofaLinné, 1758, Carcasses in Urban and Coastal Regions of São Paulo State, Brazil. Psyche: J Entomol 2015. Article ID 983980, 12 pages, Hindawi Publishing Corporation
Chakraborty A, Ghosh S, Ansar W, Banerjee D (2015a) Developmental analysis of immature stages of Sarcophaga (Parasarcophaga) albiceps Meigen, 1826 (Diptera: Sarcophagidae) on Gallus gallus carcass: Their applications as forensic indicators. IOSR J Agric Vet Sci 8(8):79–89
Chakraborty A, Ghosh S, Ansar W, Banerjee D (2015b) Thermophilic pupal endoparasitoids: Brachymeria minuta (Hymenoptera: Chalicididae) on forensic indicator Sarcophaga (Parasarcophaga) albiceps. Prommalia III:85–94
Chakraborty A, Naskar A, Parui P, Banerjee D (2016a) Developmental variation of Indian thermophilic variety of Scuttle Fly Megaselia (Megaselia) scalaris (Loew, 1866) (Diptera: Phoridae) on different substrates. Scientifica 2016. 9 pages, Article ID 4257081
Chakraborty A, Saha GK, Banerjee D (2016b) Developmental variation of two different variety of Indian blow flies: Chrysomya megacephala (Fabricius, 1794) and Lucilia cuprina (Wiedemann, 1830) (Diptera: Calliphoridae) on dead Gallus gallus (Linnaeus, 1758). J Entomol Zool Stud 4(5):881–889
Dadour I, Cook D, Fissioli JN, Bailey WJ (2001) Forensic entomology: Application, education and research in Western Australia. Forensic Sci Int 120:48–52
D'Almeida JM, Lopes HS (1983) Sinantropia de dı́pteroscaliptratos (Calliphoridae) no Estado do Rio de Janeiro. Arquivos da Universidade Federal do Rural do Rio de Janeiro 6:39–48
de Carvalho LML (2010) Toxicology and forensic entomology. In: Amendt J, Campobasso CP, Goff LM, Grassberger M (eds) Current concepts in forensic entomology. Springer, London, pp 163–178
De Jong GD (1995) Report of Chrysomya megacephala (Diptera: Calliphoridae) in northern New Mexico. Entomol News 106:192
Dufek MI, Oscherov EB, Damborsky MP, Mulieri PR (2016) Assessment of the abundance and diversity of Calliphoridae and Sarcophagidae (Diptera) in sites with different degrees of human impact in the Iberá Wetlands (Argentina). J Med Entomol 53(4):827–835
Easton AM, Smith KGV (1970) The entomology of the cadaver. Med Sci Law 10(4):208–215
Edwards CA, Reichle DE, Crossley DA Jr (1970) The role of soil invertebrates in turnover of organic matter and nutrients. In: Analysis of temperate forest ecosystems, vol 1. Springer, Berlin, Heidelberg, New York, pp 147–172
Erzinclioglu Z (1989) Entomology and the forensic scientist: how insects can solve crimes. J Biol Educ 23(4):300–302
Erzinclioglu Z (2000) Maggots, murder and men: memories and reflections of a forensic entomologist. Harley Books, Colchester
Feener DH Jr, Brown BV (1997) Diptera as parasitoids. Annu Rev Entomol 42:73–97
Galante E, Marcos-Garcia MA (2008) Decomposer insects. In: Capinera JL (ed) Encyclopedia of entomology. Springer, Dordrecht
Gennard DE (2007) Forensic entomology: an introduction. John Wiley & Sons, Ltd., Chichester

Goff ML (2000) A fly for the prosecution—how insect evidence helps solve crimes. Harvard University Press, Cambridge, MA

Greenberg B (1991) Flies as forensic indicators. J Med Entomol 28:565–577

Greenberg B, Kunich JC (2002) Entomology and the law: flies as forensic indictors. Cambridge University Press, Cambridge, p 306

Gruner SV, Slone DH, Capinera JL (2007) Forensically important Calliphoridae (Diptera) associated with pig carrion in rural north-central Florida. J Med Entomol 44:509–515

Grzywacz A, Hall MJ, Pape T, Szpila K (2017) Muscidae (Diptera) of forensic importance—an identification key to third instar larvae of the western Palaearctic region and a catalogue of the muscid carrion community. Int J Leg Med 131(3):855–866

Hall RD (1990) Medicocriminal entomology. In: Catts EP, Haskell NH (eds) Entomology and death: a procedural guide. Joyce's Print Shop, Clemson, SC

Hall RD (2001) Introduction: perceptions and status of forensic entomology. In: Byrd JH, Castner JL (eds) Forensic entomology: the utility of arthropods in legal investigations. CRC Press, Boca Raton, FL, pp 1–15

Hall MJR, Brown T, Jones P, Clark D (2008) Forensic sciences. In: Cox M, Flavel A, Hanson I, Laver J, Wessling R (eds) The scientific investigation of mass graves: towards protocols and standard operating procedures. Cambridge University Press, Cambridge, pp 463–497

Hall M, Whitaker A, Richards C (2012) Chapter 8: forensic entomology. In: Grant NM, Roberts J (eds) Forensic entomology hand book: from crime science to court, 1st edn. John Wiley and Sons Ltd

Hanski I (1987) Carrion fly community dynamics: patchiness, seasonality and coexistence. Ecol Entomol 12:257–266

Hart AJ, Whitaker AP, Hall MJR (2008) The use of forensic entomology in criminal investigations: how it can be of benefit to SIOs. J Homicide Major Incid Invest 4:37–47

Haskell NH, Williams RE (eds) (2008) Entomology and death: a procedural guide. Joyce's Print Shop, Clemson, SC

Henssge C, Madea B, Gallenkemper E (1988 Oct) Death time estimation in case work. II. Integration of different methods. Forensic Sci Int 39(1):77–87

Heo CC, Mohamad AM, John J et al (2008) Insect succession on a decomposing piglet carcass placed in a man-made fresh-water pond in Malaysia. Trop Biomed 25:23–29

Hore G (2021) Studies on necrophagous flies (Insecta: Diptera) of forensic importance in Urban and semi-urban areas of Kolkata and adjoining areas. Ph.D. Thesis,. University of Calcutta, Kolkata, India

Hore G, Banerjee D (2017) Necrophagous flies (Insecta: Diptera) and their role in maintaining ecosystem balance. ENVIS Newsletter (Theme-Ecosystem services), ENVIS Centre on Faunal Diversity. Zool Surv India 23(3):2–6

Hore G, Maity A, Naskar A, Ansar W, Ghosh S, Saha GK, Banerjee D (2017a) Scanning electron microscopic studies on antenna of *Hemipyrellia ligurriens* (Wiedemann,1830) (Diptera: Calliphoridae)—a blow fly species of forensic importance. Acta Trop 172:20–28

Hore G, Parui P, Saha GK, Banerjee D (2017b) Variations in colonization and succession pattern of dipteran flies of forensic importance on Indian mole-rat carcasses in urban and suburban localities of Kolkata, West Bengal: implications in corpse relocation studies. Malaysian J Med Res 1(3):52–62

Hore G, Saha GK, Banerjee D (2018) Sensory organs of forensically important fly *Ophyra capensis* (Wiedemann,1818) (Diptera: Muscidae): a scanning electron microscopic study. Acta Trop 185: 400–411

Joseph ANT, Parui P (1980) Filth inhabiting flies (Diptera) of Calcutta city. Bull ZoolSurv India 3(1 & 2):1–12

Joseph I, Mathew DG, Sathyan P, Vargheese G (2011) The use of insects in forensic investigations: an overview on the scope of forensic entomology. J Forensic Dent Sci 3(2):89–91. https://doi.org/10.4103/0975-1475.92154

Kashyap VK, Pillai VV (1989) Efficacy of entomological method in estimation of post mortem interval: a comparative analysis. Forensic SciInt 40:245–250

Kitchell JF, O'Neill RV, Webb D, Gallep GW, Bartell SM, Koonce JF, Ausmus BS (1979) Consumer regulation of nutrient cycling. Bioscience 29:28–34

Kumara TK, Abu Hassan A, Che Salmah MR, Bhupinder S (2009) The infestation of Dermestes ater (De Geer) on a human corpse in Malaysia. Trop Biomed 26:73–79

Kuusela S, Hanski I (1982) The structure of carrion fly communities: the size and the type of carrion. Holarctic Ecol 5:337–348

LeBlanc HN, Logan JG (2010) Exploiting insect olfaction in forensic entomology. In: Amendt J, Goff ML, Campobasso CP, Grassberger M (eds) Current concepts in forensic entomology, vol 2. Springer, Netherlands, pp 205–221

Lord WD, Stevenson JR (1986) American registered professional entomologists. Chesapeake Chapter, Washington DC

Majer JD (1987) Invertebrates as indicators for management. In: Nature conservation: the role of remnants of native vegetation. Surrey Beatty and Sons Pty Limited with CSIRO and CALM, New South Wales, Australia, pp 353–354

Majumdar SK, Jana I, Misra KK (2007) Synanthropy of Carrion flies in three districts of Southern West Bengal, India. Int J Ecol Environ Sci 33(1):29–39

Matuszewski S, Bajerlein D, Konwerski S, Szpila K (2010) Insect succession and carrion decomposition in selected forests of Central Europe. Part 2: Composition and residency patterns of carrion fauna. Forensic Sci Int 195(1-3):42–51

Megnin P (1894) La Faune des Cadavres. Encyclopedie Scientifique des Aide-Memoire. G. Mas-son and Gauthier-Villars et Fils, Paris

Midgley JM, Richards CS, Villet MH (2010) The utility of Coleoptera in forensic investigations. In: Amendt J, Goff ML, Campobasso CP, Grassberger M (eds) Current concepts in forensic entomology. Springer, Dordrecht, pp 57–68

Mondor EB, Tremblay MN, Tomberlin JK, Benbow EM, Tarone AM, Crippen TL (2012) The ecology of Carrion decomposition. Nat Educ Knowl 3(10):21

Moore JC, Berlow EL, Coleman DC, De Ruiter PC, Dong Q, Hastings A, Johnson NC, Mccann KS, Melville K, Morin PJ, Nadelhoffer K, Rosemond AD, Post DM, Sabo JL, Scow KM, Vanni MJ, Wall DH (2004) Detritus, trophic dynamics and biodiversity. Ecol Lett 7:584–600

Nandi BC (2002) auna of India and the adjacent countries-Diptera: Sarcophagidae (vol. X), vol i–xxiv. Zoological Survey of India, Kolkata, pp 1–608

Ndueze OU, Okiwelu SN, Umeozor OC, Noutcha MAE (2013) Arthropod succession on wildlife carcasses in lowland rainforest, Rivers State, Nigeria. Eur J Exp Biol 3(5):106–110

Odat N, Hasan HS, Obeidat M, Aladaileh S (2015) Relationships between species diversity and evenness of necrophagousDiptera and environmental conditions in three habitats of Jordan. J Entomol Zool Studies 3(5):89–94

Pape T (1996) Catalogue of the Sarcophagidae of the world (Insecta: Diptera). Mem Entomol Inter 8:1–558. Florida, Gainesville: Associated Publishers

Pereira de Sousa JR, Esposito MC, Carvalho-Filho FDS, Juen L (2014) The potential uses of sarcosaprophagous flesh flies and blowflies for the evaluation of the regeneration and conservation of forest clearings: a case study in the Amazon forest. J Insect Sci 14:215

Pereira de Sousa JR, Carvalho-Filho FDS, Esposito MC (2015) Distribution and Abundance of Necrophagous Flies (Diptera:Calliphoridae and Sarcophagidae) in Maranhao, Northeastern Brazil. J Insect Sci 15(1):70

Pereira de Sousa JR, Carvalho-Filho FDS, Juen L, Esposito MC (2016) Evaluating the effects of different vegetation types on necrophagous fly communities (Diptera: Calliphoridae; Sarcophagidae): implications for conservation. PLoS One 11(10):e0164826

Povolny D, Verves YG (1997) The flesh-flies of central Europe (Insecta, Diptera, Sarcophagidae). Spixiana Suppl 24:1–260

Ren L, Yanjie S, Wei C, Fanming M, Jifeng C, Guanghui Z, Lushi C, Yong W, Jianqiang D, Yadong G (2018) A brief review of forensically important flesh flies (Diptera: Sarcophagidae). Forensic Sci Res 3(1):16–26

Ricklefs R (1996) Invitación a la Ecología. In: La economía de la naturaleza. Editorial MédicaPanamericana

Rochefort S, Giroux M, Savage J, Wheeler TA (2015) Key to forensically important Piophilidae (Diptera) in the Nearctic region. Can J Arthropod Identif 27. https://doi.org/10.3752/cjai.2015.27

Roeterdink EM, Dadour IR, Watling RJ (2004) Extraction of gunshot residues from the larvae of the forensically important blowfly Calliphora dubia (Macquart) (Diptera: Calliphoridae). Int J Leg Med 118:63–70

Samuel WM (1988) The use of age classes of winter ticks on moose to determine time of death. Can Soc Forens Sci J 21:54–59

Seastedt TR, Crossley DA Jr (1984) The influence of arthropods on ecosystems. Bioscience 34(3): 157–161

Silahuddin SA, Latif B, Kurahashi H et al (2015) The Importance of habitat in the ecology of decomposition on rabbit carcasses in Malaysia: implications in forensic entomology. J Med Entomol 52:9–23

Singh D, Bharti M (2008) Some notes on the nocturnal larviposition by two species of Sarcophaga (Diptera: Sarcophagidae). Forensic Sci Int 177:19–20

Singh N, Kumari V, Singh NP (2016) An analysis of the ecological succession pattern of Diptera on the carcass of laboratory bred rats. Int J Agric Innov Res 4(5):2319–1473

Sinha S.K. (2009) Sarcophagidae, Calliphoridae and Muscidae (Diptera) of the Sundarbans Biosphere Reserve, West Bengal, India. Rec Zool Surv India Occ. Paper No. 308, pp. 1–46 (Director, Zool. Surv. India, Kolkata).

Smith KGV (1986) A manual of forensic entomology. British Museum (Natural History) and Cornell University

Šuláková H (2014) Forenzní entomologie – když smrt je začátek. Živa 5:250–256

Swift B (2010) Methods of time since death estimation within the early post-mortem interval. J Homicide Major Incid Invest 6:97–112

Szpila K, Madra A, Jarmusz M et al (2015) Flesh flies (Diptera: Sarcophagidae) colonising large carcasses in central Europe. Parasitol Res 114:2341–2348

Thyssen PJ (2009) Keys for identification of immature insects. In: Current concepts in forensic entomology. Springer, Dordrecht, pp 25–42

Tomberlin JK, Benbow ME (2015) Forensic entomology international dimensions and frontiers. CRC Press, Boca Raton, FL

Ubelaker DH (2013) Postmortem interval. In: Siegel JA, Saukko PJ, Houck MM (eds) Encyclopedia of forensic sciences, 2nd edn. Academic Press, pp 24–27

Wells JD, LaMotte LR (2001) Estimating the post-mortem interval. In: Byrd JH, Castner JL (eds) Forensic entomology: the utility of arthropods in legal investigations. CRC, Boca Raton, FL. WHO experts consultation on Ebola Reston pathogenicity in humans. Geneva, Switzerland 1 April 2009. WHO/HSE/EPR/2009.2:25

Yang LH, Gratton C (2014) Insects as drivers of ecosystem processes. Curr Opin Insect Sci 2:26–32

Forensic Flies: Advances in Study Through Morphological Evidences

3

Abstract

Microscopy is by far the most powerful tool in identification and study of the forensic flies. Our knowledge on fly morphology, reproductive parts, genitalia, male accessory glands, developmental stages of eggs (immature, pupa, larva), adult stages, salivary glands, antennal sensilla, and alimentary canals of flies are largely based on studies from light and scanning electron microscope that enables their identification, and some features like male genitalia find importance as identification keys in Calliphoridae. SEM studies have enabled identification of *Chrysomya* species from immature forms from corpses. In this chapter we highlight the advances of morphology-based studies in flies of forensic importance.

Keywords

Light microscopy (LM) · Scanning electron microscopy (SEM) · Forensic flies · Diptera · Beetles

Abbreviations

C. albifrontalis	*Calliphora albifrontalis*
C. dubia	*Calliphora dubia*
C. graham	*Calliphora graham*
C. macellaria	*Cochliomyia macellaria*
C. megacephala	*Chrysomya megacephala*
C. putoria	*Chrysomya putoria*
C. rufifacies	*Chrysomya rufifacies*
C. stygia	*Calliphora stygia*
C. tibialis	*Conicera tibialis*

S. Ghosh, D. Banerjee, *Biology of Forensically Important Invertebrates*,
https://doi.org/10.1007/978-981-99-5026-3_3

C. varipes	*Chrysomya varipes*
C. vicina	*Calliphora vicina*
C. vomitoria	*Calliphora vomitoria*
E. peregrinum	*Euryomma peregrinum*
E. philippinensis	*Exoristobia philippinensis*
F. albitarsis	*Fannia albitarsis*
F. femoralis	*Fannia femoralis*
F. fusconotata	*Fannia fusconotata*
F. heydenii	*Fannia heydenii*
F. sanihue	*Fannia sanihue*
H. aenescens	*Hydrotaea aenescens*
H. ignava	*Hydrotaea ignava*
H. illucens	*Hermetia illucens*
H. ligurriens	*Hemipyrellia ligurriens*
H. rostrate	*Hydrotaea rostrate*
H. similis	*Hydrotaea similis*
H. violacea	*Hypopygiopsis violacea*
L. ampullacea	*Lucilia ampullacea*
L. caesar	*Lucilia caesar*
L. cuprina	*Lucilia cuprina*
L. eximia	*Lucilia eximia*
L. illustris	*Lucilia illustris*
L. sericata	*Lucilia sericata*
L. silvarum	*Lucilia silvarum*
LM	Light microscopy
M. domestica	*Musca domestica*
O. obesa	*Ornidia obesa*
P. regina	*Phormia regina*
P. terraenovae	*Protophormia terraenovae*
P. vulgaris	*Parapiophila vulgaris*
S. albiceps	*Sarcophaga albiceps*
S. caerulescens	*Sarcophaga caerulescens*
S. dux	*Sarcophaga dux*
S. impatiens	*Sarcophaga impatiens*
S. martellata	*Sarcophaga martellata*
S. nudiseta	*Synthesiomyia nudiseta*
S. peregrine	*Sarcophaga peregrina*
S. ruficornis	*Sarcophaga ruficornis*
SEM	Scanning electron microscopy

3.1 Introduction

Forensic entomology is the science of application of insects to investigate crime and deaths and uses the study of insects in forensic analysis.

Insects and other arthropods that colonize and grow on and near a dead decaying body collected, identified, and studied are crucial in estimating the postmortem interval (PMI). The species growing on the dead body and the detection of the life cycle stage indicates the time of its colonization and hints at the time elapsed since the death. This is vital information to estimate the time of death of humans and animals. Therefore, in order to estimate PMI accurately, it is very important to estimate and evaluate the insects at site of crime. This requires a careful and proper examination of the dead body and its environment, studying of patterns of succession of insects. Forensic entomology thus finds importance in forensic investigations in medicolegal, criminal investigations, detection of drugs/poisons used for the murder, etc., finding the incident location and estimation of the time of wound infliction. In this chapter, we highlight the studies on morphology of the Dipteran flies of forensic importance.

3.2 Evolution of Forensic Entomology

The physician from Italy, Francesco Redi, as early as in 1668, proposed the theory of spontaneous generation. The prevalent theory at that times was of spontaneous generation of maggots from a rotting meat piece. However, Redi found that maggots develop requirement exposure of meat to air either fully or partially. This led to revolution in forensic entomology highlighting the fact that that maggots failed to develop on rotting meat that was un-exposed to air. This led to understanding of the fact that the insect life cycle played a major role in forensic entomology.

Hermann Reinhard, a German doctor, exhumed corpses and studied the development of diverse insect species on buried bodies. He also found that not all insects found underground on a buried body were associated with the body. In the nineteenth century, the physician from France, Dr. Louis François Etienne Bergeret, pioneered the study of forensic entomology and its application in PMI estimation. Jean Pierre Mégnin, a French veterinarian, found out the insect succession on exposed versus buried corpses (Benecke 2008).

3.3 Flies in Forensic Entomology

Flies are the first insects attracted to a decaying corpse. They start a life cycle, and the immature forms, larva, pupa, and maggots, feed on moist fluids of the cadavers. The flies studied most from the point of view of forensic entomology are as follows.

Fly	Features
Blow flies	• Metallic like texture of body • Can smell and sense dead body from around 16 km distance
Flesh flies	• Breed on decaying matter, corpses, garbage, dung, etc. • Some lay eggs on exposed wounds • Viviparous in nature
House flies	• Comon in occurrence • Disease carriers
Cheese flies	• Occur in animal products, excreta, fungi, etc. • Larvae infest decaying animal bodies, smoked fish, cured meat, cheese, milk products • Larval form crucial for estimating TOD (time of death)
Beetles	• Attracted to dead bodies late when compared to flies • Moth flies replace beetles in drier conditions
Dermestid beetles	• commonly called as skin or hide beetles • Infest dead bodies after soft tissues devoured by other organisms • Thrive on hair or skin • Commonly found insects from human corpses
Bone beetles	• Collected at late decomposed stage of corpse. • Under family Cleridae • Thrive on decaying flesh
Carrion beetles	• Larval form thrive on vertebrate carcasses • Adults eat maggots
Hide beetles	• Thrive on decaying carcasses and animal products like dried fish, bacon, dog treats, cheese/milk products • Occur in clusters around food for feeding and mating
Rove beetles	• Feed on cadavers or carcasses • Survive on maggots and other insect larvae on carrions
Scarab beetles	• Members of Scarabaeidae family • includes dung beetles • commonly occur under carrion and corpses • Dung, fungi, and decaying flesh are favored source of food • Play role as clean-ups
Sap beetles	• Frequently occurring near fermenting or decaying plant fluids, rotting fruits, tree sap • Some infest carcasses
Clown beetles	• Commonly called hister beetles • Prefer hiding under carcass at day times, emerge at night and, feed on maggots, larvae, etc.
Dung beetles	• Carron feeders, decaying fungi, manure, etc.
Mites	• *Macrocheles* mites feed on carrions at early decomposed condition of dead body • Tyroglyphidae under family Acaridae and Oribatida mites (Acari: Oribatida) survive on dry skin at later point
Moths	• Larval form active during twilight and daytime • Feed on the corpse hair • Found at last stages of decomposition

Insects and arthropod specimens at all stages of decomposition should be collected from open spaces, wounds and other body parts and surfaces, cloth folds, soil in which dead body lay, carpet, curtain, etc. in the dead body environment. Insects are found from wounds and in and around natural orifices of a dead body. Dipterans and Coleopteran beetles are the frequent ones to be found.

Both types of insect look very different at different stages of their lives.

3.3.1 Flies (Diptera) (Table 3.1)

Eggs	• Tiny, usually in masses • Found in a wound or natural orifice, on clothing, etc. • Paint brush dipped in water or forceps used for collection • 75% alcohol or 50% isopropyl alcohol used for preservation of half of the collection • Other half may be stored in vial without getting dehydrated • Live eggs need to be observed every few hours and time of hatching and the approximate percentage can be recorded
Maggots	• Range from 1–2 mm to 17 mm in size • Found crawling and may occur in masses • The maggot masses are heat generating, which expedites development • Temperature and size of each mass may be recorded • Large maggots are older • Samples from different areas of the body to be collected
Larva	• Third instar larvae in search of a suitable place to pupate leave the food source • Due to their wandering nature, the area around the dead body should be scanned to search and collect them • Due to the burrowing nature under leaf litter, soil below the corpse should be scanned • For dead body lying on a slope, the insect may feed on body fluids seeping downhill and hence the trail of fluid needs to be scanned to search and collect them • Corpse, clothes, and surroundings should be searched
Pupae and/or empty pupal cases	• Needs careful search • Occur on cloth, hair, or soil around a dead body • Chose dry areas to pupate • May hide in cuffs, pockets, seams, etc. • Clothing, boxes, curtains rugs, etc. around a dead body should be searched • 2–20 mm in size, oval, dark brown in color • An empty pupal case open at one end, through which adult fly has emerged
Adults	• Indicating the species of insect • Adult with crumpled wings indicates just emergence • Needs to properly collected and stored • May be collected by net or inverted vial • Empty pupal cases indicate adult fly emerged marking completion of one generation of flies of this species

Table 3.1 Morphology and identification of all the developmental stages of forensic flies[a]

Developmental stages	Morphology and techniques used	
	LM	SEM
Eggs	1. 1% $KMnO_4$ solution/1 min for, dehydration in 15, 70, and 95%, absolute alcohol (1 min/solution) and permanent mounting 2. Measure of egg mean length, width of median area 3. Staining of hatching pleats, width of plastron, morphology of plastron area surrounding the micropyle (Sanit et al. 2013) 4. Chorionic sculpturing (Sukontason et al. 2004)	1. Eggs were collected, fixed in alcoholic Bouin, prepared for SEM 2. Mean length, median area, and width of eggs measured 3. Hatching pleats, structure of chorion, body, appearance of eggs observed (Sanit et al. 2013) 4. Anastomosis or holes at the top of the islands (Mendonça et al. 2008) 5. Chorionic sculpturing, width of plastron (Sukontason et al. 2007)
Larvae (maggots)	1. Mean body length of all instar larvae is measured 2. Body appearance, cephalopharyngeal skeleton, dorsal cuticular spines between the prothorax and mesothorax, feature of the posterior spiracle, pseudocephalon, antennal complex, maxillary palpus, facial mask, cephaloskeleton, thoracic and abdominal spinulation, spiracular field and posterior spiracles observed (Szpila et al. 2013)	1. Anterior and posterior spiracles, cephalopharyngeal skeleton, characteristics of the dorsal spines between the prothorax and mesothorax of all instar larvae studied 2. Quantitation of number of sensory papillae on the anterior spiracle, oral grooves, and posterior spiracular hairs 3. Pseudocephalon, antennal complex, maxillary palpus, facial mask, cephaloskeleton, thoracic and abdominal spinulation, spiracular field and posterior spiracles studied (Szpila et al. 2013)
Pupa	1. Clearing technique to pale the integument of fly puparia, for visible observation of second to fourth segments (anterior end), posterior spiracle and oral grooves (Sukontason et al. 2007) 2. Measurement of length and width of puparia, posterior spiracular hairs length, width of puparia, number of papillae in the anterior spiracle, length of the gaps between the posterior spiracles (Sukontason et al. 2007)	1. Observation of cuticular sculpture of tubercles along the dorsal and lateral segments; bubble membrane on the dorsolateral border of the fifth segment; the morphology of integument; presence of spines on integument; structure of the posterior spiracle, its hair; oral grooves and pupal respiratory horns (Sukontason et al. 2006a, b, c)
Winged adult	1. Body appearance and color 2. Measurement of female size 3. Quantitation of the number of live larvae carried, time taken to larviposit live larvae, mean length and width of fly 4. Morphology, measurement of male internal reproductive organs,	1. Observing the differences in antennal sensilla, adhesive device (pulvilli) between the tarsal claws of the legs 2. Measurement of mean length of adult fly 3. Differences in antennal and maxillary palps of male and female

(continued)

Table 3.1 (continued)

Developmental stages	Morphology and techniques used	
	LM	SEM
	spermatozoa, and spermiogenesis (Name et al. 2012) 5. Shape of the intersegmental spines between the pro- and mesothorax, posterior spiracles, number of papillae on the anterior spiracles, oral grooves, and posterior spiracular hairs (Sukontason et al. 2006a, b, c)	4. Trichoid, basiconic, coeloconic, styloconic, and sensory pit in antennal scape, pedicel, and flagellum and in maxillary palps 5. Morphology, measurement of male internal reproductive organs, spermatozoa, and spermiogenesis (Name et al. 2012) 6. Shape of the intersegmental spines between the pro- and mesothorax posterior spiracles, number of papillae on the anterior spiracles, oral grooves, and posterior spiracular hairs (Sukontason et al. 2006a, b, c)

[a]*LM* light microscopy, *SEM* scanning electron microscopy. Sample preparation for study under LM and SEM reveal differences. Table adapted with permission from Ghosh et al. (2018)

3.3.2 Beetles

1. Occur in adult, larvae, pupae.
2. All stages find importance.
3. Capable of fast movement.
4. Appear often underneath the body, or under clothing.
5. Collected in vials separately and stored with some air.
6. They may be preserved in alcohol.

3.3.3 Other Insects

Other insects like cheese skippers may be present. Insects observed, colonizing diverse parts of the body, are collected and kept separately. Different species should be kept separately as most specimens are fragile and they are to be collected with care. Small or delicate specimens can be collected from dead body, clothes, folds of dress, and nearby surfaces of dead body by an fine artist's brush either dipped in water or alcohol and kept in well-sealed vials. The insects should be preserved by entomologist and studied.

Important observations on habitat like locality collected from places like woods, beach, house, and roadside area; adjoining vegetation like bush, trees, fallen leaves, grass, shrubs, etc.; soil type collected like rocky, sandy, muddy, etc. are to be recorded. The weather parameters like temperature during collection time, sunny or cloudy day, humidity at time of collection, elevation of the site, and coordinates of map at the site of death are to be recorded along with the specimens collected. The

Table 3.2 Limitations of forensic entomology

Factor	Limitation
Temperature of the death site	• Very important factor • Far-away record of weather from site of crime • Differences in microclimate of the corpse and that of the surrounding area more so find importance for maggot mass • Presence of maggot masses finds importance
Seasonal effect	• Seasonal nature of forensic entomology like in Canada • Used in spring, summer, and fall with insect abundance • Used less in winter, with no or very few insects
Time of results	• Not immediate • Due to time taken for rearing insects • DNA evidence used to speed up immature stage identification
Body disposed of in a way without insects	(a) Freezing—a frozen body being placed outside at a later date may give wrong impression about death incident date (b) Burial—in deeply buried bodies, most insects are excluded (c) Insects may be excluded in wrapped or packed dead bodies
Drugs	Drugs/toxins affect insect development
Insect succession	Insect succession shows variation with geographic region

site where the dead body lay like under shade or direct sunlight, anything unusual observed, may be recorded (Table 3.2).

Other uses for insects in forensic science	Hints at translocation of the body from the scene
	Indicative of any disturbance caused to dead body post-crime, thereby causing disturbance in insect life cycle
	Detection of wound site and decomposition by detecting the specific pattern, of insect colonization
	Presence of drugs or toxins
	Used to place a suspect at the crime, insect inside a cocklebur weed has been used to identify a rapist
	Use of insect evidence in civil cases as well
	Could indicate child or senior abuse/neglect

Despite being effective in crime detection, detection of PMI forensic entomology suffers from certain limitations

3.4 Morphology and Studies of Different Flies of Forensic Importance

Since accurate identification of forensic flies collected from dead bodies studies finds utmost importance in the forensic studies, it is therefore important to understand and study the insects properly and characterize them accurately. Several studies have been conducted across these lines across the globe. In this chapter we present an account of different morphology-based studies on different flies of forensic importance (Table 3.3).

3.5 SEM Studies on Forensic Flies

Blow flies male genitalia of flies under subfamilies of Chrysomyinae including *Ch. bezziana Villeneuve, Ch. rufifacies (Macquart), Ch. chani Kurahashi, Ch. pinguis (Walker), Ch. thanomthini Kurahashi and Tumrasvin, Ch. nigripes Aubertin, Ch. villeneuvi Patton* and *Hypopygiopsis tumrasvini Kurahashi, Hypopygiopsis infumata (Bigot), Luciliinae, Hemipyrellia ligurriens (Wiedemann), Lucilia (L) including L. papuensis Macquart, L. cuprina (Wiedemann),* and *L. sinensis Aubertin L. porphyrina (Walker)* from Thailand studied by SEM revealed distinguishing cercus features and surstylus in shapes and distiphallus as structurally complex morphology (Sontigun et al. 2018).

Sarcophagidae flies studied under LM, image-stacking stereomicroscopy, and SEM revealed differences in sclerotization level of parts of the cephaloskeleton, posterior spiracular peritreme and posterior spiracular slit shape (Szpila et al. 2015).

Hermetia illucens Linnaeus or black soldier fly under Stratiomyidae, of both sexes, were studied for on maxillary palps by SEM, revealing morphology and microtrichia and sensilla distribution. The maxillary palps, with two segments, in length and shape revealed sexual dimorphism. First segment in both sexes revealed covered by microtrichia, while the second one comprised of coverage by microtrachia proximally and distally containing both microtrichia and sensory structures. Sensory pits (two types) and chaetic sensilla (one type) were studied. Female palp size with higher number of sensory pits reveals sexual dimorphism (Pezzi et al. 2021). A detailed description has been drawn in 2017 (Pezzi et al. 2017).

Eggs, first, second, third instar larvae, puparium of *Musca domestica* Linnaeus, 1758, studied by SEM revealed eggs with similarity in all muscid flies. The exochorion revealed modified cells used for respiration by embryo. The first and second instar larvae revealed openings (02 in number) on the posterior spiracle, and a last one revealed anterior spiracle. The third instar larvae, in some other Muscidae, revealed posterior spiracular opening with a sinuous near the spiracular plate edge. The puparium revealed similar morphology to third instar with a respiratory structure that plays a role in respiration during metamorphosis (Cortinhas et al. 2020).

Puparia of *Boettcherisca (B)* including *B. nathani Lopes, 1961*; *B. peregrina Robineau-Desvoidy, 1830*; *Liopygia ruficornis Fabricius, 1794*; *Lioproctia pattoni Senior-White, 1924*; and *Parasarcophaga (Liosarcophaga) dux Thomson, 1869* studied by SEM revealed differences between species in arrangement of papillae in the anterior spiracle and its number and intersegmental spine shape between the prothorax, mesothorax, and the spiracular tufts pattern at the posterior spiracle (Samerjai et al. 2014).

Ch. megacephala, Ch. putoria, L. eximia L. cuprina, and *Ophyra aenescens*, using SEM, reveal lack of anastomosis or holes at island; *Ch. megacephala* and *Ch. putoria* revealed few anastomoses but no holes; *L. eximia* and *O. aenescens* bore

Table 3.3 Important species of forensic entomology identified based on morphological approaches[a]

Species names	Morphological method
Lucilia cuprina (Wiedemann 1830) Australian sheep blow fly)	• SEM ultrastructure used to identify morphology of immature stages from China (Sukontason et al. 2004) • Ultrastructure of sensilla, mouthparts, and antennae of male and female flies (Hassan et al. 2013) • Morphology of male internal reproductive organs, spermatozoa, and spermiogenesis studied using LM and TEM (Name et al. 2012)
Lucilia ampullacea (Villeneuve, 1922) *Lucilia caesar* (Linnaeus 1758) *Lucilia illustris* (Meigen 1826)	Antennal complex, pseudocephalon, maxillary palpus, facial mask, cephaloskeleton, thoracic and abdominal spinulation, spiracular field, and posterior spiracles of first instars of the European and Mediterranean blow flies studied by SEM and LM (Szpila et al. 2013)
Lucilia sericata (Meigen 1826)	• Studied on dead porcine and bovine tissues as bait (Clark et al. 2006) Pseudocephalon, antennal complex, maxillary palpus, facial mask, cephaloskeleton, thoracic and abdominal spinulation, spiracular field, and posterior spiracles of first instars of the European and Mediterranean blow flies studied by SEM and LM (Szpila et al. 2013) • Third instar larvae with diagnostic features from closely related species (Erzinclioglu 1987)
Lucilia silvarum (Meigen 1826)	• Adults from corpse at 3350 m altitude reported from rocky mountains in June (Adair and Kondratieff 2006). Pseudocephalon, antennal complex, maxillary palpus, facial mask, cephaloskeleton, thoracic and abdominal spinulation, spiracular field, and posterior spiracles of first instars of the European and Mediterranean blow flies studied by SEM and LM (Szpila et al. 2013)
Lucilia eximia (Wiedemann 1819)	• SEM of eggs and the immature stages in China (Mendonca et al. 2012a, b, c) • Internal reproductive organs, spermatozoa, and spermiogenesis in males studied by LM and TEM (Name et al. 2012)
Calliphora graham (Aldrich 1930)	• Spines of the body segments of the third instar larva studied LM (Velásquez et al. 2010).
Calliphora vomitoria (Linnaeus 1758)	• Studied on dead porcine and bovine tissues as bait (Clark et al. 2006) Pseudocephalon, antennal complex, maxillary palpus, facial mask, cephaloskeleton, thoracic and abdominal spinulation, spiracular field, and posterior spiracles of first instars of the European and Mediterranean blow flies studied by SEM and LM (Szpila et al. 2013)
Calliphora vicina (Robineau-Desvoidy 1830)	• Pseudocephalon, antenna, maxillary palpus, facial mask, labial lobe, thoracic and abdominal spinulation, spiracular field, posterior spiracles, anal pad, and cephaloskeleton of first instar larva were identified by SEM LM and illustrations (Szpila et al. 2008, 2013).

(continued)

Table 3.3 (continued)

Species names	Morphological method
Calliphora stygia (Fabricius 1781)	• Third instar larvae studied by LM (Erzinclioglu 1987).
Calliphora dubia (Macquart 1855)	• Female fly • Size and the number of live larvae carried were studied (Cook and Dadour 2011)
Calliphora albifrontalis (Mauoch 1932)	• Thorax of adult nonmetallic blue-black in color • Abdomen predominantly brown or brown-yellow • Studied by LM • Third-instar larvae with a cephalopharyngeal skeleton with a pigmented accessory oral sclerite (Wall and Shearer 2012)
Chrysomya megacephala (Fabricius 1794)	• Morphology and identification of fly eggs by morphometric quantitation of mean length, width of median area, and darkness staining of hatching pleats by SEM (Sanit et al. 2013)
Chrysomya putoria (Wiedemann 1830) (African latrine fly)	• SEM study of eggs and the immature stages from China (Mendonca et al. 2012a, b, c) • Third instar larvae of closely related species studied by LM (Erzinclioglu 1987)
Chrysomya rufifacies (Macquart 1843)	• Third instar larvae of closely related species studied by LM (Erzinclioglu 1987) • Morphology of fly eggs based on morphometric quantitation of mean length, width of median area, and darkness staining of hatching pleats by SEM (Sanit et al. 2013)
Chrysomya varipes (Macquart 1851)	• Secondary fly reported in possum carcass with variability in the numbers (Lang et al. 2006) • Face wholly yellow; middle and hind femora broadly reddish basally; male frons extremely broad, almost as broad as in females, male front femur mostly whitish with prominent, long whitish hairs dorsally by LM (James 1971)
Hypopygiopsis violacea (Macquart 1835)	• Cephalopharyngeal skeleton, anterior and posterior spiracles of the second and third instar larvae studied by LM (Ahmad et al. 2010)
Cochliomyia macellaria (Fabricius 1775)	• Flies, pupas, and larvae feed on deeply buried corpse (Balme et al. 2012) • Morphological identification of the spermatheca by LM (Name et al. 2012)
Protophormia terraenovae (Robineau-Desvoidy 1830)	• Flies, pupas, and larvae feed on deeply buried corpse (Balme et al. 2012) • Pseudocephalon, antenna, maxillary palpus, facial mask, labial lobe, thoracic and abdominal spinulation, spiracular field, posterior spiracles, anal pad, and cephaloskeleton of first instar larva identified by SEM and LM (Szpila et al. 2008, 2013)
Hemipyrellia ligurriens (Wiedemann 1830)	• The morphology and developmental rate of all stages of flies (Bunchu et al. 2012; Sukontason et al. 2008) • Morphometric analysis of the length and width of

(continued)

Table 3.3 (continued)

Species names	Morphological method
	puparia, along with the length of the gaps between the posterior spiracles (Sukontason et al. 2007)
Phormia regina (Meigen 1826)	• Correlation between response of flies to temperature in colonizing carcasses and oviposition (Matuszewski et al. 2014) • Pseudocephalon, antenna, maxillary palpus, facial mask, labial lobe, thoracic and abdominal spinulation, spiracular field, posterior spiracles, anal pad, and cephaloskeleton of first instar larva studied by SEM and LM (Szpila et al. 2008)
Hydrotaea ignava (Harris 1780) *Hydrotaea similis* (Meade 1887)	• Morphology of adult flies and oviposition varies with an abrupt and large increase in a narrow range of low temperatures and no response in a broad range of high temperatures (Matuszewski et al. 2014)
Fannia fusconotata (Rondani 1868)	• The abundance of flies, time of occurrence, and residency time at the pig carcasses vary with season and environment • Grow maximally when pig carcasses were kept in the sun (Aballay et al. 2012)
Fannia albitarsis (Stein 1911)	• Identified in pig carcasses in semi-arid climate (Aballay et al. 2012)
Fannia femoralis (Stein 1896)	
Fannia heydenii (Wiedemann 1830)	
Hydrotaea aenescens (Wiedemann 1830) (black dump flies)	• Salivary gland hypertrophy not significant as compared to *Musca domestica* (Geden et al. 2011)
Hydrotaea rostrata (Robineau-Desvoidy 1830)	• Appears as a tertiary fly in possum carcass with enormous variability in the numbers (Lang et al. 2006)
Synthesiomyia nudiseta (Wulp 1883)	• Morphology and identification of eggs of different flies based on morphometric quantitation of mean length, width of median area, and darkness staining of hatching pleats by SEM (Sanit et al. 2013) • All larval instar morphology studied by LM and SEM (Velásquez et al. 2010)
Musca domestica (Linnaeus 1758) Muscidae	• Maggots (day 33) in dry decomposed monkey carcass in Malaysia (Chen et al. 2010) • Morphology and identification of fly eggs of different flies based on morphometric quantitation of mean length, width of median area, and darkness staining of hatching pleats by SEM (Sanit et al. 2013) • Morphometric analysis of the length and width of puparia, along with the length of the gaps between the posterior spiracles of seven fly species, displayed differences among them (Sukontason et al. 2007)
Sarcophaga albiceps (Meigen 1826)	• Costal spine small, sixth tergite dorsally divided (Encyclopedia of Life 2014)

(continued)

Table 3.3 (continued)

Species names	Morphological method
Sarcophaga dux (Thomson 1869) *Sarcophaga ruficornis* (Fabricius 1914)	Posterior spiracle, number of papillae on the anterior spiracle of the third instars; characteristic of adult males of Sarcophagidae (Sukontason et al. 2010)
Sarcophaga martellata (Senior-White 1924)	• Male genitalia with unambiguous species identification characteristics, but the female flies reveal similarity in general morphology (Tan et al. 2010)
Sarcophaga impatiens (Walker, 1849)	• Male genitalia and larval stages were identified (Encyclopedia of Life 2014)
Sarcophaga caerulescens (Zetterstedt 1838)	• Male abdomen with long and dense hair, female scutellum with an additional lateral bristle (Encyclopedia of Life 2014)
Sarcophaga peregrina (Robineau-Desvoidy 1830)	• Age of adult males and females detected by pteridine fluorescence analysis (Zhu et al. 2013)
Fannia sanihue (Domínguez and Aballay 2008)	• Grow maximally when pig carcasses were kept in shade (Aballay et al. 2012)
Euryomma peregrinum (Meigen 1826)	• Identified in pig carcasses in semi-arid climate (Aballay et al. 2012)
Desmometopa sp.	• Breed on indoor human carrion, human excreta, and manure with other Sarcophagidae (Kumara et al. 2010)
Conicera tibialis (Scmitz 1925) (coffin flies)	• Detected in an 18-year-old buried corpse. Adult fly, newly matured flies, and puparia were frequently found in old buried corpse (Martín-Vega et al. 2011)
Parapiophila vulgaris (Fallen 1820)	Occurrence time and activity period of some taxa insect succession in pig carrion decomposition were studied in pine-oak forest, hornbeam-oak forest, and alder forest in Europe (Matuszewski et al. 2014)
Exoristobia philippinensis (Ashmead 1904)	• Predates on pupa of *Chrysomya* sp. on monkey carcass in Malaysia (Chen et al. 2010) • Body is more or less uniform metallic black, antenna with yellowish scape, legs are yellowish-brown, femora are slightly darker, wings are hyaline. Mandibles with three nearly equal teeth. Antenna with six-segmented funicle, compound eyes with conspicuous, dense hairs. Mesoscutum with deep reticulate sculpture, scutellum with much shallower reticulate sculpture. Cercal plates situated at about middle of gaster fore wing with marginal vein longer than broad; postmarginal vein much shorter than stigmal vein (Ashmead 1904)
Hermetia illucens (Linnaeus, 1758) (black soldier fly)	• Fly and larva have been used to determine PMI of cadaver in northern Brazil (Pujol-Luz et al. 2008). • Arrangement and shape of spiracular openings, structures of the anal segment, and the cephalopharyngeal skeleton were used for the identification of third instars of the species present in the Iberian peninsula (Velásquez et al. 2010)

(continued)

Table 3.3 (continued)

Species names	Morphological method
Ornidia obesa (Fabricius 1775)	• Female genitalia segments are tawny pilose, female mesonotum is almost entirely tawny pilose, 2 male abdomen is shiny except black pollinose on most of second tergum in male (Thompson 1991)

[a]Adapted with modification with permission from Ghosh et al. (2018)
[b]*PMI* Postmortem interval, *COI* cytochrome oxidase subunit I, *COII* cytochrome oxidase subunit II, *LM* light microscopy, *SEM* scanning electron microscopy, *TEM* transmission electron microscopy

both anastomoses and holes; and *L. cuprina* bore only anastomoses. *Ch. megacephala* and *L. eximia* revealed median bifurcation, and *O. aenescens* revealed anterior end bifurcation and in *Ch. putoria* and *L. cuprina*. Being rounded. Diagnostic features include T chorionic cell sculptures, length, and way that median area ends up posteriorly (Mendonça et al. 2008). *Ch. megacephala* immature forms are known to cause myiasis. The immature forms, larvae and puparia, were studied by SEM. All larval instar body of *Ch. megacephala* revealed overall similar features with smooth integument with small spines located at limit of all segments. The cephalic region reveals robust spines groups with one or two tips. The puparia reveal similarity to third instar larvae, other than for retracted cephalic structures. The wrinkled integument and spiracle disc at posterior end with spiracular openings (03) localized on elevated tops characterizes the third instar larva (Mendonca et al. 2012a, b, c). *Ch. nigripes Aubertin* puparia found in Thailand, studied using SEM, revealed accurate identification of fly. Distinguishing features included aggregated puparia adhering side by side, tibia of a skeletonized corpse with triangular-shaped anterior end, thoracic segments (03), and broad hairy patches observed at the sixth segment dorsolaterally (Sukontason et al. 2006a, b, c, d, e).

Ophyra capensis Wiedemann, 1818, studied in depth by SEM approaches revealed sensory organs (03 in number), ocellar region, compound eye, and adult male and female antenna; the ocellar region revealed large size in females and covered with microtrichia. Compound eye with antero-frontally located ommatidia revealed large size as compared to other facets, together with sexual dimorphism of ommatidia size. Antenna revealed sensilla (five types), chaetic sensilla (two types) on the scape and pedicel, and trichoid sensilla and basiconic sensilla (two types) on the flagellum together with microtrichia. Basiconic sensilla revealed multiporous surface with olfactory function (Hore et al. 2018).

Ultrastructures of immature forms of *Hypopygiopsis infumata (Bigot)* under SEM and LM revealed widening median area throughout entire length in eggs. Vermiform in shape, creamy white in color larva with smooth integument, pseudocephalon bearing sensory structures of antennal dome, ventral organ, and maxillary palpus. The first instar larvae, appear with two tufts of cirri around dorsal side of the mouth opening while the second and third instar larvae, reveal presence of minute tubercles (06) along the periphery of the last abdominal segment. The anterior spiracle of second and third instar, and puparia reveal fan shaped, single row, of 9–11 papillae. Puparia, at first abdominal segment, reveal membrane cluster. *H. infumata* immature forms reveal similarity to other *Hypopygiopsis* species. Oral sclerite and dorsal cornua rough surface could distinguish *H. infumata* and *H. tumrasvini* (Sanit et al. 2018).

Boettcherisca nathani and *Lioproctia pattoni* larva under SEM revealed marked differences in the third instar larva in spines between the pro- and mesothorax. On the other hand, *B. nathani and L. pattoni* reveal slender triangular and stout triangular spines (Samerjai et al. 2016) respectively. While *B. nathani* has branched long, fine hairs, those in *L. pattoni* are long and fine with moderate branching. The anterior spiracle in *B. nathani* has two irregular rows of 21–27 papillae while a single irregular row of 20–28 papillae in *L. pattoni. Parasarcophaga dux, Bercaea africa*, and *Liopygia ruficornis* were studied by LM (Samerjai et al. 2016).

Eggs of *Ch. rufifacies (1.34 mm), Ch. megacephala* with swollen hatching pleats, *Ch. pinguis (1.65 mm), Ch nigripes, Hypopygiopsis tumrasvini (1.63 mm), L cuprina, L. porphyrina (145 mm)*, and *M. domestica* revealed variation in mean length of long eggs; thus, egg length, median area width, and hatching pleats in their darkness were distinctive features (Sanit et al. 2013).

L. sinensis Aubertin immature forms were studied by SEM. Eggs have slight widening median area throughout length. The terminal abdominal segment of the first instar reveals tubercles, and the surface integument bears circular rows of microtrichia. Each anterior spiracle with 9–12 papillae was arranged in a single row in the second and third instars. In all instars and puparia, the presence of prominent outer ventral tubercle is a prominent feature of *L. sinensis* capable of distinguishing them from other Lucilia species (Sanit et al. 2017).

L. cuprina causative agent of human myiasis and its immature stages were identified by using keys and SEM studies (Mendonça et al. 2014a, b).

Triceratopyga calliphoroides Rohdendorf, common in distribution in eastern Asia, with olfaction system guiding fly behavior was studied using SEM and laser scanning confocal microscopy (LSCM) by examining the sensillar morphology of adults, sensory pit, and sacculus. The presence of microtrichia and mechanoreceptors occurs in antennal scape and pedicel. Antennal sensilla (eight types) on external surface of antennal funiculus including trichoid, basiconic, coeloconic, and clavate-like sensilla were detected on sensory pits or sacculus (Zhang et al. 2014).

SEM study on antenna and maxillary palp of *Sarcophaga tibialis Macquart* adults of both sexes revealed the antenna as typical of cyclorrhaphan Diptera, with three segments, including scape, pedicel, and postpedicel, bearing the arista. The scape was capped by microtrichia with a row of long chaetic sensilla. The pedicel was enveloped by microtrichia with three types of chaetic sensilla together with cluster of setiferous plaques. Clavate, styloconic, trichoid, and basiconic sensilla occur on the postpedicel (Pezzi et al. 2016).

Peckia (Euboettcheria) collusor fly larvae can cause myiasis. In the Neotropical regions, immature stages of Sarcophagidae studied under SEM enabled their identification. Net-like patches in integument, intersegmental spines, shape, and disposition of larval instars were identifying features. Spiracular ramifications varied in number at larval instars. The puparia and third instar larvae are similar except for the retracted cephalic structures (Mendonca et al. 2013).

Immature stages of *Hemipyrellia ligurriens*, isolated from Thailand studied under SEM revealed eggs with wide plastron region spanning the whole length with upright median hatch line. There is a presence of external chorionic sculpture in a hexagonal arrangement with elevated reticular boundaries. The second and third instar larvae revealed differences in posterior spiracular slits. Body size and structure of the spiracles both at anterior and posterior end were studied. The coarctate puparium featured a clustered bubble membrane either side of the first abdominal segment replaced by shorter tubular respiratory horns, in old puparia (Sukontason et al. 2008).

Ophyra aenescens immature forms studied under SEM revealed spiracular opening of the first instar with interrupted the length at the middle, with developed ventral tubercles and anterior spiracles with variation in spiracular ramification numbers. The puparia and third instar larvae reveal similarity but with retracted cephalic region and presence of respiratory structure on the third thoracic segment (Cortinhas et al. 2016).

Morphology of the first instar larvae of *Ch. rufifacies, Ch. incisularis, Ch. nigripes, Ch. varipes, Ch. saffranea,* and *Ch. latifrons* with facial mask, pseudocephalon, maxillary palpus, antennal complex, posterior thoracic spiracles, abdominal spinulation, cephaloskeleton structures, and spiracular field was studied, and identification key of Australian *Chrysomya* was documented in 2016 (Szpila and Wallman 2016).

In Thailand, *Ch. rufifacies (Macquart)* and *Ch. villeneuvi Patton* are closely related species in larvae and puparia morphology and aggressive feeding habit of *Ch. rufifacies*. Morphology of puparia of both blow flies was studied using SEM. Cuticular sculpture of tubercles with different features were studied in both flies. Morphology of net-like patches in integument and posterior spiracle structure of both species was observed to be identical (Sukontason et al. 2006a, b, c, d, e).

Larval morphology of all three instars of *Atherigona orientalis (Schiner),* common in tropics and subtropics, was studied. The unpaired sclerite was observed in second and third instar larvae. Observation of a sensory organ X in all three larval instars representing Muscidae was reported. The bubble membrane was observed in second and third instar (Grzywacz and Pape 2014).

Chrysomya flies originally found occurring in Asia, Australia, and Africa were introduced in America as sheep parasite, in the 1970s. *Ch. megacephala* egg ultramorphology collected in Brazil revealed oval shape with one flat and another convex face, 0.52 ($\pm$0.03) mm in length and 0.12 ($\pm$0.02) mm in width with circular micropyle, with two perforations with small projections (David et al. 2008).

Ch. putoria (Wiedemann) larvae and puparia were studied using SEM. The first instar larvae with 12 smooth segments are observed separated by spines with visible antennae and maxillary palps. All three instar reveal similarities, except the presence of anterior spiracle with 11–12 spiracular ramifications. At the anal segment, the presence of two spiracular openings is observed in second instars and three openings in third instar larvae. Puparia had retracted cephalic region and no visible head structures (Mendonça et al. 2012a, b, c).

C. albiceps eggs, larvae, and puparia were studied using SEM. Eggs have elongated Y-shaped anterior and tapered posterior end. The micropyle has some projections. The first instar larva had 12 segments separated by spines, visible antennae, and maxillary palps. Second and third larval instars had revealed similarity to first instar, except anterior spiracle. Body tegument has net-like patches and tubercles and robust and erect tubercles at the third instar larvae. Puparia reveal retracted cephalic region and curved tubercles (Mendonça et al. 2010).

Ch. megacephala isolated from India studied under SEM revealed basiconic trichoid, sensilla and microtrichia on labellar lobes, and haustellum with maxillary palps of mouthparts (Bhattacherjee et al. 2022).

Piophila (P) Fallén, 1810, including two species *P. casei* and *P. megastigmata* share ecological niche and colonize advanced decay stages of human corpses. The *P. megastigmata* immature stages studied by LM and SEM revealed pseudocephalon features of antenna, facial mask, maxillary palps, etc.; anterior and posterior spiracles; tegumentary sculpturing cephalopharyngeal skeleton; and anal division. Second and third instar reveal morphological similarity, while the first instar larva reveals particular features. Shape, arrangement of oral combs, number, oral ridges, scales of spiracular field, papillae of anterior and posterior spiracles, sensilla of maxillary palpus could distinguish *P. megastigmata* and *P. casei* (Paños et al. 2013).

The immature developmental stages of Sarcophaga *(Liopygia) cultellata from* southern part of France and northern part of Italy have common ecological niche with *S. (Liopygia) argyrostoma and S. (Liopygia) crassipalpis*. Immature stages of *S. (L.) cultellata* under LM and SEM reveal that *S. (L.) crassipalpis*, *S. (L.) cultellata,* and *S. (L.) argyrostoma* could be identified separately in all immature stages by both LM and SEM. Tegumental warts, fan-shaped anterior spiracle with 15–18 respiratory papillae (in single row) in third instar larvae of *S. (L.) cultellata*, make it unique from others (Ubero-Pascal et al. 2015).

Puparia of *M. domestica L.* and *Ch. megacephala (F.),* studied under SEM, revealed morphological differences. Housefly puparia appear evenly rounded at ends with anterior spiracle bearing six papillae, lateral pupal respiratory horns (a pair) prior to posterior boundary of the first abdominal segment, and papillae with longitudinal opening. Pupa of blow fly reveals anterior spiracle of 8–12 papillae. The protruding pupal respiratory horns and 38 globules on the bubble-like membrane were observed (Siriwattanarungsee et al. 2005).

The third instar larva of a Sarcophaga, collected from a mummified body revealed similarity to other sarcophagid larvae. Specific features include number and arrangement of papillae on the anterior spiracle, spine structure, circumspiracular tubercle at caudal end, and branching of the posterior spiracular hairs (Sukontason et al. 2003a, b).

Cochliomyia macellaria or secondary screwworm is known to cause disease and transport *Dermatobia hominis* eggs, causing livestock loss. Studied using SEM, immature forms were studied. The ellipse-shaped egg has concave dorsal surface, with the islands in the median area lacking anastomosis, but with some perforations. Second instar larva and other immature forms revealed intersegmental spines, spine bands at the abdominal segments. Two spiracular openings on the first and second larval instars are characteristics of *Cochliomyia*. Spine tips of the cephalic region, the abdominal intersegmental bands, and spiracular openings are diagnostic features of *C. macellaria* (Mendonça et al. 2014a, b).

Eggs of Calliphoridae species under six genera were studied with SEM revealing variability within and among members of same species. Similarities in morphological features within congeneric species and similarities between some species of different genera were observed (Greenberg and Singh 1995).

Ravinia belforti larva by SEM reveal muscoid vermiform pattern with 12 segments, with pointed anterior end and thicker posterior end. The flat spines of the cephalic collar with double to quadruple points differ from the single-pointed spines. In second instar larva, the presence of anterior spiracle papillae (16–22) differs from other species. Posterior spiracles are observed within peritreme. Internalized spiracular cavity in the posterior region is unique for Sarcophagidae. More visible features and thick and dense developed spines over cephalic collar are observed, near the first thoracic segment in third instar larva. Puparium reveals similarity to other Sarcophagidae species (da Silva-Xavier and de Carvalho Queiroz 2016).

C nigripes Aubertin larva was studied by LM and SEM. The cephalopharyngeal skeleton, anterior and posterior spiracle, and dorsal spines lying between the pro- and mesothorax are highlighted (Sukontason et al. 2005).

Hypopygiopsis (*H. tumrasvini Kurahashi)* larvae and puparia were studied under SEM. Vermiform-shaped cream color larvae with smooth integument. The pseudocephalon has sensory structures, including antenna, maxillary palpus, and ventral organ. First instar reveals two tufts of cerri dorsal of the mouth opening, and second and third instars have strong mouth hook. Sensory structure or Keilin's organ is ventrally located on thoracic segments in all second and third instars, with six conspicuous tubercles present along periphery of the terminal abdominal segment and large puparia (7.77–9.51 mm long and 3.10–3.97 mm wide). First abdominal segment, with cluster of ~125 bubble membranes in latero-dprsal side with young puparia; but older one with a minute pupal respiratory horn (Sanit et al. 2012).

Ultrastructures of the surface of puparia of *L. cuprina* and *Liosarcophaga dux* were studied by SEM. *L. cuprina* puparium is 6.2 ± 0.2 mm long and 2.3 ± 0.1 mm wide. Anterior spiracle has 5–7 papillae in a single row. Triangular intersegmental spines between the pro- and mesothorax have constricted tips. Clustered bubble membrane in dorsolateral side, with 30 mammillate structures, on each side of the first abdominal segment in young puparia but in aged puparia short, tubular respiratory horns replaces it. The puparium posterior end is broadly truncated with a pair of posterior spiracles, medially, each with three straight spiracular slits. The *L. dux* (9.9 ± 0.3 mm length and 3.8 ± 0.2 mm width) puparium is larger than *L. cuprina*. An anterior spiracle has 14–17 papillae. Broad and triangular intersegmental spines can be observed between the pro- and mesothorax. In *L. dux,* slightly tapered terminal tail segment of the puparium is abruptly truncated with two posterior spiracles. D-shaped posterior spiracular disc with a pronounced medial projection and three long, narrow spiracular slits, vertically oriented were also observed (Sukontason et al. 2006a, b, c, d, e).

Hydrotaea similis larvae isolated from corpses were studied. The third instar larvae of both *H. dentipes* and *H. similis* were studied under SEM and LM revealing unique identifying features of the two species (Grzywacz et al. 2014).

Sarcophagid genus *Sarcophaga Meigen, 1826,* and Sarcophaga (Liosarcophaga) tibialis in the Iberian peninsula studied using LM and SEM microscopy approaches revealed cephalopharyngeal skeleton, hook-shaped mouth, external morphology and intermediate and basal sclerites and spiracles, and identification keys based on light microscopy in *S. (L.) tibialis* for every immature stage (Paños-Nicolás et al. 2015).

First instar larvae of *Ch. megacephala* morphology studied through SEM revealed similarity to *Ch rufifacies*, and relative thickness of the posterior spiracular ends enabled to differentiate developmental stage (Sukontason et al. 2003a, b).

SEM studies of spiracles, both anterior and posterior, of first and second instar larvae of *Megaselia scalaris (Loew)* enabled study of larval transformation, feeding behavior, and mode of respiration and find importance in forensic studies (Boonchu et al. 2004).

3.6 Discussion

Although morphology-based studies on forensically important insects find importance and have been highlighted in this chapter, there are certain limitations associated with no availability of appropriate keys to identify them accurately and distinguish between closely related species and groups. Thus, molecular studies have found more and more importance in identifying closely related species/immature stages of forensic flies. In the next few chapters, we discuss how the molecular approaches are enabling identification of closely related mature and immature forms.

References

Aballay FH, Domínguez MC, Fernández Campón F (2012) Adult Fanniidae associated to pig carcasses during the winter season in a semiarid environment: initial examination of their potential as complementary PMI indicators. Forensic Sci Int 219(1–3):284.e1–284.e4

Adair TW, Kondratieff BC (2006) Three species of insects collected from an adult human corpse above 3300 m in elevation: a review of a case from Colorado. J Foren Sci 51:1164–1165

Ahmad FMS, Marwi MA, Syamsa RA, Zuha RM, Ikhwan Z, Omar B (2010) Morphological descriptions of second and third instar larvae of Hypopygiopsis violacea Macquart (Diptera: Calliphoridae), a forensically important fly in Malaysia. Trop Biomed 27:134–137

Ashmead WH (1904) A list of the Hymenoptera of the Philippine Islands with descriptions of new species. J NY Entomol Soc 12:1–22

Balme GR, Denning SS, Cammack JA, Watson DW (2012) Blow flies (Diptera: Calliphoridae) survive burial: evidence of ascending vertical dispersal. Forensic Sci Int 216:e1–e4

Benecke M (2008) A brief survey of the history of forensic entomology. Acta Biol Benrodis 14:15–38

Bhattacherjee R, De S, Sharma G, Ghosh S, Mishra S, Suman DS, Banerjee D (2022) Prevalence of mouthpart sensilla and protease producing symbiotic gut bacteria in the forensic fly Chrysomya megacephala (Fabricius, 1794): insight from foraging to digestion. Acta Trop 229:106380

Boonchu N, Sukontason K, Sukontason KL, Chaiwong T, Piangjai S, Vogtsberger RC (2004) Observations on first and second-instar larvae of Megaselia scalaris (Loew) (Diptera: Phoridae). J Vector Ecol 29(1):79–83

Bunchu N, Thaipakdee C, Vitta A, Sanit S, Sukontason K, Sukontason KL (2012) Morphology and developmental rate of the blow fly, Hemipyrellia ligurriens (Diptera: Calliphoridae): forensic entomology applications. J Parasitol Res 2012:371243. https://doi.org/10.1155/2012/371243

Chen CD, Lee HL, Nazni WA, Ramli R, Jeffery J, Sofian-Azirun M (2010) First report of the house fly larvae, Musca domestica (Linnaeus) (Diptera: Muscidae) associated with the monkey carcass in Malaysia. Trop Biomed 27(2):355–359

Clark K, Evans L, Wall R (2006) Growth rates of the blowfly, Lucilia sericata, on different body tissues. Foren Sci Int 156:145–149

Cook DF, Dadour IR (2011) Larviposition in the ovoviviparous blow fly Calliphora dubia. Med Vet Entomol 25:53–57

Cortinhas LB, Mendonça PM, Barbosa RR, Queiroz MMC (2016) Ultrastructure of immature stages of the black dump fly: Ophyra aenescens (Wiedemann, 1830) (Diptera: Muscidae: Azeliinae). Acta Trop 158:125–129

Cortinhas LB, Martins Mendonça P, Braga MV, Queiroz MMC (2020) Ultrastructure of the immature stages of Musca domestica (Diptera: Muscidae: Muscinae). J Med Entomol 57(6): 1712–1721

da Silva-Xavier A, de Carvalho Queiroz MM (2016) Ultrastructure analysis of the immature stages of Ravinia belforti (Diptera: Sarcophagidae), a species of medical-veterinary and forensic importance, by scanning electron microscopy. Acta Trop 159:192–199

David JA, Rocha T, Caetano FH (2008) Ultramorphological characteristics of Chrysomya megacephala (Diptera, Calliphoridae) eggs and its eclosion. Micron 39(8):1134–1137

Encyclopedia of Life. (2014), www.gbif.org/species/5064140

Erzinclioglu YZ (1987) The larvae of some blowflies of medical and veterinary importance. Med Vet Entomol 1(2):121–125

Geden CJ, Steenberg T, Lietze VU, Boucias DG (2011) Salivary gland hypertrophy virus of house flies in Denmark: prevalence, host range, and comparison with a Florida isolate. J Vector Ecol 36(2):231–238

Ghosh S, Ansar W, Banerjee D (2018) Diagnosis of crime reporter flies in forensic entomology: a review. Indian J Entomol 80(2):158–176

Greenberg B, Singh D (1995) Species identification of calliphorid (Diptera) eggs. J Med Entomol 32(1):21–26

Grzywacz A, Pape T (2014) Larval morphology of Atherigona orientalis (Schiner) (Diptera: Muscidae): a species of sanitary and forensic importance. Acta Trop 137:174–184

Grzywacz A, Lindström A, Hall MJR (2014) Hydrotaea similis Meade (Diptera: Muscidae) newly reported from a human cadaver: a case report and larval morphology. Forensic Sci Int 242:e34–e43

Hassan MI, Fouda MA, Hammad KM, Basiouny AL, Kamel MR (2013) The ultrastructure of sensilla associated with mouthparts and antennae of Lucilia cuprina. J Egypt Soc Parasitol 43(3): 777–785

Hore G, Saha GK, Banerjee D (2018) Sensory organs of forensically important fly Ophyra capensis (Wiedemann, 1818) (Diptera: Muscidae): a scanning electron microscopic study. Acta Trop 185:400–411

James MT (1971) Two new species of Phaenicia from the West Indies (Diptera: Calliphoridae). Proc Entomol Soc Wash 73:381–385

Kumara TK, Abu Hassan A, Che Salmah MR, Bhupinder S (2010) A report on the pupae of Desmometopa sp.(Diptera: Milichiidae) recovered from a human corpse in Malaysia. Trop Biomed 27:131–133

Lang MD, Allen GR, Horton BJ (2006) Blowfly succession from possum (Trichosurus vulpecula) carrion in a sheep-farming zone. Med Vet Entomol 20:445–452

Martín-Vega D, Gómez-Gómez A, Baz A (2011) The "coffin fly" Conicera tibialis (Diptera: Phoridae) breeding on buried human remains after a post-mortem interval of 18 years. J Foren Sci 56:1654–1656

Matuszewski S, Szafałowicz M, Grzywacz A (2014) Temperature-dependent appearance of forensically useful flies on carcasses. Int J Leg Med 128(6):1013–1020

Mendonça PM, dos Santos-Mallet JR, de Mello RP, Gomes L, de Carvalho Queiroz MM (2008) Identification of fly eggs using scanning electron microscopy for forensic investigations. Micron 39(7):802–807

Mendonça PM, Dos Santos-Mallet JR, Queiroz MM (2010) Ultramorphological characteristics of immature stages of Chrysomya albiceps (Wiedemann 1819) (Diptera: Calliphoridae), a fly specie of forensic importance. Microsc Res Tech 73(8):779–784

Mendonça PM, dos Santos-Mallet JR, De Carvalho Queiroz MM (2012a) Ultrastructure of larvae and puparia of the blowfly Chrysomya megacephala (Diptera: Calliphoridae). Microsc Res Tech 75(7):935–939

Mendonça PM, Dos Santos-Mallet JR, De Carvalho Queiroz MM (2012b) Ultrastructure of immature stages of the blowfly Chrysomya putoria (Wiedemann, 1818) (Diptera: Calliphoridae). Microsc Res Tech 75(2):206–211

Mendonça PM, Dos Santos-Mallet JR, De Carvalho Queiroz MM (2012c) Ultrastructure of immature stages of the blowfly Chrysomya putoria (Wiedemann, 1818) (Diptera: Calliphoridae). Microsc Res Tech 75:206–211

Mendonça PM, Barbosa RR, Cortinhas LB, Santos-Mallet JR, de Carvalho Queiroz MM (2013) Ultrastructure of immature stages of Peckia (Euboetcheria) collusor (Diptera: Sarcophagidae). Acta Trop 128(3):522–527

Mendonça PM, Barbosa RR, Carriço C, Cortinhas LB, dos Santos-Mallet JR, Queiroz MM (2014a) Ultrastructure of immature stages of Lucilia cuprina (Diptera: Calliphoridae) using scanning electron microscopy. Acta Trop 136:123–128

Mendonça PM, Barbosa RR, Cortinhas LB, dos Santos-Mallet JR, de Carvalho Queiroz MM (2014b) Ultrastructure of immature stages of Cochliomyia macellaria (Diptera: Calliphoridae), a fly of medical and veterinary importance. Parasitol Res 113(10):3675–3683

Name KP, Barros-Cordeiro KB, Filho JB, Wolff M, Pujol-Luz JR, Báo SN (2012) Structure and ultrastructure of spermatozoa and spermiogenesis in three species of Lucilia Robineau-desvoidy, 1830 (Diptera: Calliphoridae). J Morphol 273(2):160–172

Paños A, Arnaldos MI, García MD, Ubero-Pascal N (2013) Ultrastructure of preimaginal stages of Piophila megastigmata McAlpine, 1978 (Diptera, Piophilidae): a fly of forensic importance. Parasitol Res 112(11):3771–3788

Paños-Nicolás Á, Arnaldos MI, García MD, Ubero-Pascal N (2015) Sarcophaga (Liosarcophaga) tibialis Macquart 1851 (Diptera: Sarcophagidae): micromorphology of preimaginal stages of a fly of medical and veterinary interest. Parasitol Res 114(11):4031–4050

Pezzi M, Whitmore D, Chicca M, Semeraro B, Brighi F, Leis M (2016) Ultrastructural morphology of the antenna and maxillary palp of Sarcophaga tibialis (Diptera: Sarcophagidae). J Med Entomol 53(4):807–814

Pezzi M, Leis M, Chicca M, Falabella P, Salvia R, Scala A, Whitmore D (2017) Morphology of the antenna of Hermetia illucens (Diptera: Stratiomyidae): an ultrastructural investigation. J Med Entomol 54(4):925–933

Pezzi M, Scapoli C, Bharti M, Faucheux MJ, Chicca M, Leis M, Marchetti MG, Mamolini E, Salvia R, Falabella P, Bonacci T (2021) Fine structure of maxillary palps in adults of Hermetia illucens (Diptera: Stratiomyidae). J Med Entomol 58(2):658–665

Pujol-Luz JR, Francez PA, Ururahy-Rodrigues A, Constantino R (2008) The black soldier-fly, Hermetia illucens (Diptera, Stratiomyidae), used to estimate the postmortem interval in a casein Amapá state, Brazil. J Foren Sci 53:476–478

Samerjai C, Sanit S, Sukontason K, Klong-Klaew T, Kurahashi H, Tomberlin JK, Morakote N, Wannasan A, Sukontason KL (2014) Morphology of puparia of flesh flies in Thailand. Trop Biomed 31(2):351–361

Samerjai C, Sanit S, Sukontason K, Morakote N, Wannasan A, Pereira RM, Sukontason KL (2016) Morphology of immature stages of flesh flies, Boettcherisca nathani and Lioproctia pattoni (Diptera: Sarcophagidae). Acta Trop 163:109–120

Sanit S, Sukontason KL, Sribanditmongkol P, Klong-Klaew T, Samerjai C, Sontigun N, Limsopatham K, Sukontason K (2012) Surface ultrastrucure of larva and puparia of blow fly Hypopygiopsis tumrasvini Kurahashi (Diptera: Calliphoridae). Parasitol Res 111(6):2235–2240

Sanit S, Sribanditmongkol P, Sukontason KL, Moophayak K, Klong-Klaew T, Yasanga T, Sukontason K (2013) Morphology and identification of fly eggs: application in forensic entomology. Trop Biomed 30(2):325–337

Sanit S, Sukontason K, Kurahashi H, Tomberlin JK, Wannasan A, Kraisittipanit R, Sukontason KL (2017) Morphology of immature stages of blow fly, Lucilia sinensis Aubertin (Diptera: Calliphoridae), a potential species of forensic importance. Acta Trop 176:395–401

Sanit S, Limsopatham K, Klong-Klaew T, Samerjai C, Yasanga T, Sukontason K, Tomberlin JK, Sukontason KL (2018) Morphology of immature blow fly Hypopygiopsis infumata (bigot) (Diptera: Calliphoridae), a potential species of forensic importance. Acta Trop 188:168–179

Siriwattanarungsee S, Sukontason KL, Kuntalue B, Piangjai S, Olson JK, Sukontason K (2005) Morphology of the puparia of the housefly, Musca domestica (Diptera: Muscidae) and blowfly, Chrysomya megacephala (Diptera: Calliphoridae). Parasitol Res 96(3):166–170

Sontigun N, Sanit S, Wannasan A, Sukontason K, Amendt J, Yasanga T, Sukontason KL (2018) Ultrastructure of male genitalia of blow flies (Diptera: Calliphoridae) of forensic importance. Acta Trop 179:61–80

Sukontason K, Sukontason KL, Piangjai S (2003a) Scanning electron microscopy of third-instar sarcophagid (Diptera: Sarcophagidae) recovered from a mummified human corpse in Thailand. Rev Inst Med Trop Sao Paulo 45(2):95–98

Sukontason KL, Sukontason K, Piangjai S, Boonchu N, Chaiwong T, Vogtsberger RC, Kuntalue B, Thijuk N, Olson JK (2003b) Larval morphology of Chrysomya megacephala (Fabricius) (Diptera: Calliphoridae) using scanning electron microscopy. J Vector Ecol 28(1):47–52

Sukontason K, Sukontason KL, Piangjai S, Boonchu N, Kurahashi H, Hope M, Olson JK (2004) Identification of forensically important fly eggs using a potassium permanganate staining technique. Micron 35:391–395

Sukontason KL, Vogtsberger RC, Boonchu N, Chaiwong T, Sripakdee D, Ngern-Klun R, Piangjai S, Sukontason K (2005) Larval morphology of Chrysomya nigripes (Diptera: Calliphoridae), a fly species of forensic importance. J Med Entomol 42(3):233–240

Sukontason KL, Kanchai C, Piangjai S, Boonsriwong W, Bunchu N, Sripakdee D, Chaiwong T, Kuntalue B, Siriwattanarungsee S, Sukontason K (2006a) Morphological observation of puparia of Chrysomya nigripes (Diptera: Calliphoridae) from human corpse. Forensic Sci Int 161(1): 15–19

Sukontason KL, Narongchai P, Kanchai C, Vichairat K, Piangjai S, Boonsriwong W, Bunchu N, Sripakdee D, Chaiwong T, Kuntalue B, Siriwattanarungsee S, Sukontason K (2006b) Morphological comparison between Chrysomya rufifacies (Macquart) and Chrysomya villeneuvi Patton (Diptera: Calliphoridae) puparia, forensically important blow flies. Forensic Sci Int 164(2–3): 230–234

Sukontason KL, Piangjai S, Bunchu N, Chaiwong T, Sripakdee D, Boonsriwong W, Vogtsberger RC, Sukontason K (2006c) Surface ultrastructure of the puparia of the blow fly, Lucilia cuprina (Diptera: Calliphoridae), and flesh fly, Liosarcophaga dux (Diptera: Sarcophagidae). Parasitol Res 98(5):482–487

Sukontason KL, Narongchai P, Kanchai C, Vichairat K, Piangjai S, Boonsriwong W, Bunchu N, Sripakdee D, Chaiwong T, Kuntalue B, Siriwattanarungsee S, Sukontason K (2006d) Morphological comparison between Chrysomya rufifacies (Macquart) and Chrysomya villeneuvi Patton (Diptera: Calliphoridae) puparia, forensically important blow flies. Foren Sci Int 164:230–234

Sukontason KL, Boonsriwong W, Siriwattanarungsee S, Piangjai S, Sukontason K (2006e) Morphology of puparia of Megaselia scalaris (Diptera: Phoridae), a fly species of medical and forensic importance. Parasitol Res 98:268–272

Sukontason KL, Ngern-Klun R, Sripakdee D, Sukontason K (2007) Identifying fly puparia by clearing technique: application to forensic entomology. Parasitol Res 101:1407–1416

Sukontason KL, Sribanditmongkol P, Chaiwong T, Vogtsberger RC, Piangjai S, Sukontason K (2008) Morphology of immature stages of Hemipyrellia ligurriens (Wiedemann) (Diptera: Calliphoridae) for use in forensic entomology applications. Parasitol Res 103(4):877–887

Sukontason K, Bunchu N, Chaiwong T, Moophayak K, Sukontason KL (2010) Forensically important flesh fly species in Thailand: morphology and developmental rate. Parasitol Res 106(5):1055–1064

Szpila K, Wallman JF (2016) Morphology and identification of first instar larvae of Australian blowflies of the genus Chrysomya of forensic importance. Acta Trop 162:146–154

Szpila K, Pape T, Rusinek A (2008) Morphology of the first instar of Calliphora vicina, Phormia regina and Lucilia illustris (Diptera, Calliphoridae). Med Vet Entomol 22(1):16–25

Szpila K, Hall MJ, Pape T, Grzywacz A (2013) Morphology and identification of first instars of the European and Mediterranean blow flies of forensic importance. Part II. Luciliinae. Med Vet Entomol 27(4):349–366

Szpila K, Richet R, Pape T (2015) Third instar larvae of flesh flies (Diptera: Sarcophagidae) of forensic importance--critical review of characters and key for European species. Parasitol Res 114(6):2279–2289

Tan SH, Mohd Aris E, Kurahashi H, Mohamed Z (2010) A new record of Iranihindia martellata (senior-white, 1924) (Diptera: Sarcophagidae) from peninsular Malaysia and female identification using both morphology and DNA based approaches. Trop Biomed 27:287–293

Thompson CF (1991) The flower fly genus Ornidia (Diptera: Syrphidae). Proc Entomol Soc Wash 93:248–260

Ubero-Pascal N, Paños Á, García MD, Presa JJ, Torres B, Arnaldos MI (2015) Micromorphology of immature stages of Sarcophaga (Liopygia) cultellata Pandellé, 1896 (Diptera: Sarcophagidae), a forensically important fly. Microsc Res Tech 78(2):148–172

Velásquez Y, Magaña C, Martínez-Sánchez A, Rojo S (2010) Diptera of forensic importance in the Iberian Peninsula: larval identification key. Med Vet Entomol 24(3):293–308

Wall R, Shearer D (2012) Veterinary entomology: arthropod ectoparasites of veterinary importance. Springer, Netherlands

Zhang D, Liu XH, Wang QK, Li K (2014) Sensilla on the antenna of blow fly, Triceratopyga calliphoroides Rohdendorf (Diptera: Calliphoridae). Parasitol Res 113(7):2577–2586

Zhu GH, Ye GY, Li K, Hu C, Xu XH (2013) Determining the age of adult flesh flies, Boettcherisca peregrina, using pteridine fluorescence. Med Vet Entomol 27(1):59–63

Parasitoids and Forensic Flies

4

Abstract

Parasitoids are known to grow on blow flies of forensic importance on carcass environment and may bear impact to the blow flies and also play a role in estimation of the postmortem interval (PMI). Thus, the study of parasitoids, their growth, lifecycle, emergence, number, host species ranges, etc. finds importance together with the climate conditions, temperature, humidity, annual and seasonal climatic variations, food source, etc. together with host-parasitoid interaction in the carcass environment. In this chapter we have tried to present an account of the different aspects of parasitoid and their development on necrophagous blow flies and impact of climate and other parameters on them from diverse studies reported across the globe.

Keywords

Parasitoids · Necrophagous flies

Abbreviations

A. manducator Panzer	*Alysia manducator Panzer:*
B. dorsalis	*Bactrocera dorsalis*
B. minuta Linnaeus	*Brachymeria minuta Linnaeus*
C. albifrontalis Malloch	*Calliphora albifrontalis Malloch*
C. dubia Macquart	*Calliphora dubia Macquart*
Ch albiceps	*Chrysomya albiceps*
Ch. megacephala	*Chrysomya megacephala*
Ch. rufifacies	*Chrysomya rufifacies*
E. philippinensis Ashmead	*Exoristobia philippinensis Ashmead*
H. rostrata Robineau-Desvoidy	*Hydrotaea rostrata Robineau-Desvoidy*

S. Ghosh, D. Banerjee, *Biology of Forensically Important Invertebrates*,
https://doi.org/10.1007/978-981-99-5026-3_4

L. cuprina	*Lucilla cuprina*
M. domestica	*Musca domestica (L.)*
M. fascicularis Raffles	*Macaca fascicularis Raffles*
N vitripennis	*Nasonia vitripennis*
P. lacrymosa	*Ptomaphila lacrymosa*
P. terraenovae	*Protophormia terraenovae*
PMI	Post mortem *interval*
PMImin	Postmortem *interval minimum*
S cameroni	*Spalangia cameroni*
S chontalensis	*Spalangia chontalensis*
S. (Parasarcophaga) albiceps Meigen	*Sarcophaga (Parasarcophaga) albiceps Meigen*
S. bullata	*Sarcophaga bullata*
T. arnobia	*Thyrinteina arnobia*
T. zealandicus Ashmead	*Tachinaephagus zealandicus Ashmead*

4.1 Introduction

Parasitoids including wasps under Hymenoptera or files under Diptera include insects whose larvae develop on or within other host insect bodies and finally kill the host. Parasitoids are abundant in most terrestrial ecosystems and play dominating roles in population ecology by regulating and host population control and have application in forestry and agriculture as biological control agents.

Parasitoids are capable of surviving in complex environment using chemical and auditory, visual, tactile, and thermal sensation and information for host location. Host feces, volatile secretions from salivary or mandibular gland, silk, etc. can attract some parasitoids. Vibration and sometime thermal gradients are used by parasitoids feeding on concealed material. While some parasitoids attack the host, those feeding on plants orientate themselves toward greenery. Volatile chemicals in host Drosophila's microhabitat attract the parasitoids, and they feed on fermenting substrates. Some hosts show symbiotic relationships with other organisms. Wood-boring sawflies show symbiosis with fungi digesting wood. The fungus-released chemicals attract the parasitoids.

Some hosts act as if frozen in presence of parasitoid. Although hosts need to be silent, at times hosts need to advertise themselves for other reasons and thereby attract searching parasitoids. Hosts camouflage and show immobility that confer them protection against parasitoids and predators.

Bark beetles belonging to Family Scolytidae, in order to overcome tree defenses enable mass attacks of individuals, by releasing aggregation pheromone, that attract both beetles and their larval parasitoids.

Parasitoids are sensitive to their sex pheromone released by host (Godfray 2007). Necrophagous flies are of economic, health, veterinary, and forensic importance.

Parasitoids find importance as tools in controlling Dipterans. In a study from Brazil, on different adult flies belonging to the Calliphoridae, Fanniidae, Muscidae, and Sarcophagidae, parasitoids of eight species, including *Nasonia (N) vitripennis Walker (Pteromalidae), Spalangia cameroni, Spalangia chontalensis,* and *Tachinaephagus (T) zealandicus Ashmead (Encryptidae),* were reported (Kalyanasundaram and Kamala 2021).

4.2 Parasitoids and Forensic Flies: A Global Scenario

In a study from Argentina on larvae of flies under Calliphoridae, Sarcophagidae, and Muscidae, they were reported to attract parasitoids, under Braconidae, Ichneumonidae, and Pteromalidae (Hymenoptera: Parasitica). The Hymenoptera ectoparasite, *Nasonia (N) vitripennis Walker* is known to attack the puparia of blow flies, flesh flies, and house flies, which has been reported as the most abundant species with a wide range of host (Sereno et al. 2016). Another study from pig bait revealed that pig bait attracted necrophagous flies under Calliphoridae, Muscidae, and Sarcophagidae that in turn attracted parasitoids of Pteromalidae, Braconidae, and Ichneumonidae under Hymenoptera: Parasitica species. However, seasonal effects of climate bear importance in determining the number and composition of species growing in the carcasses (Horenstein and Salvo 2012). Development of *N. vitripennis* was studied on Sarcophagid host *Boettcherisca peregrina Robineau-Desvoidy* at temperatures ranging from 16 °C to 34 °C, with threshold temperature at 11.52 °C. The total developmental duration of parasitoids from egg to adult revealed variation with temperature with no larval emergence at 34 °C (Zhang et al. 2019).

Hymenopteran parasitoids infecting carrion-breeding flies surveyed for a couple of years in Western Australia revealed pronounced effects of host, effects of the seasons, and parasitism rates in regulating parasitoid-host interactions in carcass environment. *N. vitripennis, T. zealandicus, Spilomicrus* sp. *(Diapriidae),* and *Aphaereta* sp. *(Braconidae)* were reported with *N. vitripennis* and *T. zealandicus* revealing predominance but *Spilomicrus* sp. and *Aphaereta* sp. revealed intermittent occurrence. The host forensic fly species were observed to be *Hydrotaea rostrata Robineau-Desvoidy* under Muscidae and *Calliphora (C) dubia Macquart* and *Calliphora albifrontalis Malloch* under Calliphoridae (Voss et al. 2009a, b) (Fig. 4.1).

As detection of forensic fly finds importance in estimation of PMI, and since parasitic insects like ectoparasitoids Aleochara rove beetles growing on the pupae of Cyclorrhapha flies can impose altered development of immature flies, thereby indirectly affecting the PMImin. Aleochara (A) species, including *A. asiatica and A. nigra*, acted as parasitoid to necrophagous flies including *Hemipyrellia ligurriens, L. cuprina, Ch. rufifacies, Ch. megacephala,* and *Sarcophagid* sp. and non-necrophagous flies including *Bactrocera dorsalis (B. dorsalis,* Lin and Shiao 2013).

Sarcophaga bullata (S. bullata) genome has been studied for protein coding genes playing important roles in regulating their physiology, development, tolerance

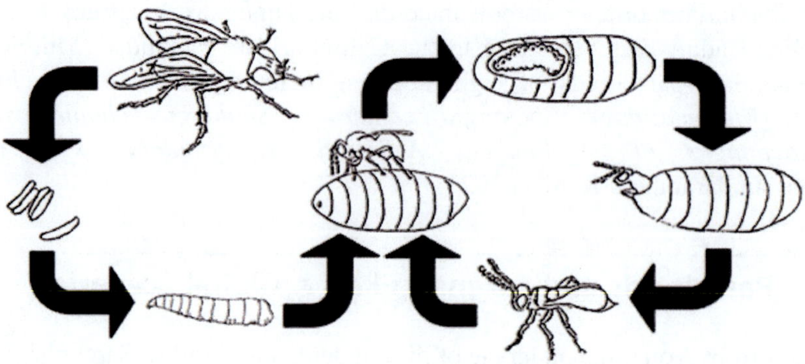

Fig. 4.1 Diagrammatic view of pteromalid pupal parasitoid of filth flies (life cycle, Image adapted from Machtinger et al. (2015) with permission)

of stress, insect diapause, neurobiology, and host-parasitoid interactions (Martinson et al. 2019).

Hymenoptera parasitoids of the fly larvae and pupae find relevance in estimating time of death (TOD) as their time of attack is often restricted to a short span of time in host insect development, but they may interfere with the developmental times of Diptera. Adult Hymenoptera parasitoids developed on decaying pig carcasses in summer revealing the presence of *N. vitripennis, T. zealandicus, Aspilota fuscicornis Haliday (Braconidae), Alysia manducator Panzer, Figites* sp. *(Figitidae), Trichopria* sp. *(Diapriidae)*, and while in the lab *N. vitripennis, T. zealandicus, Trichopria* sp., *Figites* sp., and *A. manducator* emerged from flies of Calliphoridae, Muscidae, Fanniidae, and Sarcophagidae families. Phoridae species has been reported to harbor parasite *Braconidae* (Disney and Munk 2004).

Change in the resource, carrion, ambient temperature, and season has revealed to impact the richness and community composition of insect species. This could lead to positive effects like on beetle species necrophages and predator/parasitoids impacting the species composition and richness (Dawson et al. 2021).

Pupal survival of *Chrysoma (Ch.) albiceps, Ch. megacephala,* and *Ch. putoria* and the parasitoid *N. vitripennis* experimentally submerged pupae and control ones, revealing differences in survival and emergence time indicating the real scenario of development and emergence in submerged dead bodies (Reigada et al. 2011).

Studies revealed that temperature and host species impacted T. *zealandicus* development. *T. zealandicus* revealed successful development 15–27 °C on Calliphoridae (*C.*) *albifrontalis Malloch, C. dubia Macquart, L. sericata Meigen, Ch. rufifacies Macquart*, and *Ch. megacephala Fabricius* with no parasite development at 30 °C. Temperature in host species was reported to impact development time, successful emergency and size of progeny. Significantly longer development on *Ch. megacephala* and *Ch. rufifacies* at 18–24 °C and *Ch. rufifacies and C. albifrontalis* at 15 °C and 27 °C was reported. Parasitoids successfully emerged at 21 °C declining at 15 °C and 27 °C and were significantly lower on *Ch. rufifacies* and *Ch. megacephala* (Voss et al. 2010).

Habitat of host and subsequent location encompass the foraging behavior of a parasitoid. Habitats, substrate, moisture, and light levels can affect the parasitoid host searching. Studies revealed that *N vitripennis* could penetrate and kill fly pupae located on the surface (Frederickx et al. 2014).

Year-long, seasonal, and shorter-term variation impacted patterns of insect succession onto decomposing remains from Western Australia with reported observation of *Coleoptera, Helea castor (Tenebrionidae), Omorgus tatei (Trogidae)*, and *Ptomaphila lacrymosa (Silphidae)*. The hymenopteran parasitoids on Diptera colonizers hold importance in PMI estimation (Voss et al. 2009a, b).

Microhymenopterans parasitoids of flies are reported for their forensic importance. Their parasitic habit altering post-embryonic development can alter PMI estimation. *N vitripennis* association with *Ch. megacephala* pupae revealed prolonged duration of parasitoid development, with increased female parasitoids and increased number of associated hosts. With increased number of female parasitoids per host, the possibility of increasing super-parasitism could lead to food source competition, prolonging immature development period. Increasing the number of associated hosts could lead to female distribution of their postures in available pupae, thereby causing reduced number of eggs/host and delaying the progeny development (Mello and Aguiar-Coelho 2009).

Monkey carcasses *(Macaca fascicularis Raffles)* in Malaysia in May to August 2008 attracted *Ch. rufifacies (Marquart)*, and from one pupa microhymenopteran *Exoristobia philippinensis Ashmead (Hymenoptera: Encyrtidae)* emerged, within a week (Chin et al. 2009).

T. zealandicus adults were collected from two species of Calliphoridae *Phormia regina Meigen* and *Muscidae Hydrotaea aenescens Wiedemann*, from decomposed pig, *Sus scrofa L. (Artiodactyla: Suidae)*, carcass in summer, in south central Pennsylvania (Rivers 2016) which revealed common occurrence of *T. zealandicus* in different necrophagous fly hosts.

Insect succession studies from dead European rabbit carcass in Kingdom of Saudi Arabia (KSA) revealed that it attracted Calliphorid flies including *C. vicina Robineau-Desvoidy, Ch. marginalis, Ch. chloropyga, Ch. albiceps*, and *L. cuprina*, the sarcophagid *Liosarcophaga babiyari (Lehrer)*, Muscoid flies *M. domestica (L.)* and *M. stabulans Fallén*, and Coleoptera, including dermestid Dermestes maculatus (De Geer), hister beetles, *Saprinus ruber gemminigeri*, and *Saprinus chalcites (Illiger), Saprinus moyses Marseul, Hymatismus* sp., *Necrobia ruficollis (F.)*, and *Creophilus maxillosus (L.)* beetle. Pupal parasitoid *Nasonia* sp. (Abouzied 2014) was also observed.

Parasitic wasp *N. vitripennis* development from oviposition to pupation, followed by development from oviposition to adult, was studied in the laboratory at temperatures of 15–35 °C using host pupae of the Calliphoridae blow fly *Protophormia terraenovae (Robineau-Desvoidy)* with total development time from oviposition to adult emergence ranging from 43 to 11 days reared at different temperatures of 15–30 °C, respectively, with no development at 35 °C.

The parasitoid infestation of forensic dipterans by *Lucilia* sp. and *Sarcophaga* sp. by *Alysia ridibunda* (1829) under Hymenoptera: Braconidae on beef substrate

Fig. 4.2 *B. minuta Linnaeus, 1767* from pupa of *Sarcophaga (Parasarcophaga) albiceps Meigen*, 1826 from India (Chakraborty et al. (2015), reproduced with permission from Director Zoological Survey of India, Kolkata)

was reported by Lindquist (1940) and Olivia (2008). Effects of temperature on parasitoids in Brazil were reported by Dr. De Almeida et al. (2002).

Pupal endoparasitoid *Brachymeria minuta Linnaeus, 1767* under Hymenoptera,: Chalicididae was reported from pupa of Sarcophaga *(Parasarcophaga) albiceps Meigen*, 1826 from India, using chicken bait studied under different weather conditions all-round the year (Chakraborty et al. 2015). Very recently, *Brachymeria minuta* has been reported for the first time from Brazil and America, being a parasite on pupa of the eucalypt defoliator Lepidopteran Geometridae *(Thyrinteina arnobia*, Pinheiro et al. 2023) (Figs. 4.2 and 4.3).

4.3 Discussion

Not many studies appear in this vital domain of forensic fly and the parasitoid, infection, life cycle, development, interaction, and emergence. Based on existing reserach the process appears to be dynamic. However, the host (fly)-parasitoid interaction within the carcass environment vs. studies in controlled laboratory environment is yet to be done to highlight the host-parasitoid interaction and how they may alter estimation of PMI under different climatic conditions. More research dedicated to such parasitoid-forensic fly interaction is the very need of the hour to understand the PMI estimation in a more precise and accurate way. As the parasitoids hold importance as biological control agents, the study of host (fly) interaction with the parasitoid finds importance at this point of time.

Fig. 4.3 Lateral (**a**) and dorsal (**b**) view of the parasitoid *B. minuta* from *Thyrinteina arnobia* pupa. Image reproduced from Pinheiro et al. (2023), under open access, CC BY licence

References

Abouzied EM (2014) Insect colonization and succession on rabbit carcasses in Southwestern Mountains of the Kingdom of Saudi Arabia. J Med Entomol 51(6):1168–1174

Cárcamo MC, Guidotti L, Abreu TF, Egewarth J, Duarte JP, Costa V (2021) First list of parasitoids in flies of medical, veterinary, and forensic importance in Rio Grande do Sul—Brazil. Braz J Biol 10(82):e238772

Chakraborty A, Ghosh S, Ansar W, Banerjee D (2015) Thermophilic pupal endoparasitoids: Brachymeria minuta (Hymenoptera: Chalicididae) on forensic indicator Sarcophaga (Parasarcophaga) albiceps. Prommalia III:85–94

Chin HC, Ahmad NW, Lim LH, Jeffery J, Omar B, Dhang CC, Weng LK, Sofian-Azirun M (2009) Predation on pupa of *Chrysomya rufifacies* (Marquart) (Diptera: Calliphoridae) by parasitoid, Exoristobia philippinensis Ashmead (Hymenoptera: Encyrtidae) and Ophyra spinigera larva (Diptera: Muscidae). Trop Biomed 26(3):369–372

Dawson BM, Wallman JF, Evans MJ, Barton PS (2021) Is resource change a useful predictor of carrion insect succession on pigs and humans? J Med Entomol 58(6):2228–2235

De Almeida MAF, Do Prado AP, Geden CJ (2002) Influence of temperature on development time and longevity of Tachinaephagus zealandicus (Hymenoptera: Encyrtidae), and effects of nutrition and emergence order on longevity. Environ Entomol 31(2):375–380

Disney RH, Munk T (2004) Potential use of Braconidae (Hymenoptera) in forensic cases. Med Vet Entomol 18(4):442–444

Frederickx C, Dekeirsschieter J, Verheggen FJ, Haubruge E (2014) Depth and type of substrate influence the ability of Nasonia vitripennis to locate a host. J Insect Sci 14:58

Godfray HCJ (2007) Parasitoids. In: Encyclopedia of biodiversity. Elsevier, Oxford

Horenstein MB, Salvo A (2012) Community dynamics of carrion flies and their parasitoids in experimental carcasses in Central Argentina. J Insect Sci 12:8

Lin SW, Shiao SF (2013) Life history data on the fly parasitoids Aleochara nigra Kraatz and A. asiatica Kraatz (Coleoptera: Staphylinidae), and their potential application in forensic entomology. Forensic Sci Int 232(1–3):46–55

Lindquist AW (1940) The introduction of an indigenous blowfly parasite, Alysia ridibunda say, into Uvalde County, Texas. Ann Entomol Soc Am 33:103–112

Martinson EO, Peyton J, Kelkar YD, Jennings EC, Benoit JB, Werren JH, Denlinger DL (2019) Genome and ontogenetic-based transcriptomic analyses of the flesh fly, Sarcophaga bullata. G3 (Bethesda) 9(5):1313–1320

Mello RDS, Aguiar-Coelho VM (2009) Durations of immature stage development period of Nasonia vitripennis (Walker) (Hymenoptera: Pteromalidae) under laboratory conditions: implications for forensic entomology. Parasitol Res 104(2):411–418

Olivia A (2008) Parasitoid wasp (Hymenoptera) from puparia of Sarco saprophagus flies (Diptera: Calliphoridae and Sacrophagidae) in Buenos Aires, Argentina. Rev Soc Entomol Argent 67(3–4):139–141

Pinheiro RA, Zanuncio JC, Manickavasagam S, Lemes PG, Serrão JE, Sores MA, Zanetti R (2023) Brachymeria minuta (Hymenoptera: Chalcididae): parasitism on pupa of the eucalypt defoliator Thyrinteina arnobia (lepidoptera: Geometridae) and its first record in the Americas. Brazilian J Biol 83:e262945

Reigada C, Gião JZ, Galindo LA, Godoy WA (2011) Survival of submerged blowfly species and their parasitoids: implications for postmortem submersion interval. Forensic Sci Int 212(1–3): 126–129

Rivers DB (2016) First record of the larval endoparasitoid Tachinaephagus zealandicus (Hymenoptera: Encyrtidae) from fly hosts developing on burnt remains and in Pennsylvania. J Med Entomol 53(3):732–735

Sereno AP, Salvo A, Battán-Horenstein M (2016) Trophic interactions between parasitoids and necrophagous flies in Central Argentina. Acta Trop 162:229–232

Voss SC, Spafford H, Dadour IR (2009a) Annual and seasonal patterns of insect succession on decomposing remains at two locations in Western Australia. Forensic Sci Int 193(1–3):26–36

Voss SC, Spafford H, Dadour IR (2009b) Hymenopteran parasitoids of forensic importance: host associations, seasonality, and prevalence of parasitoids of carrion flies in Western Australia. J Med Entomol 46(5):1210–1219

Voss SC, Spafford H, Dadour IR (2010) Temperature-dependent development of the parasitoid Tachinaephagus zealandicus on five forensically important carrion fly species. Med Vet Entomol 24(2):189–198

Zhang Y, Wang Y, Liu C, Wang J, Hu G, Wang M, Yang L, Chu J (2019) Development of Nasonia vitripennis (Hymenoptera: Pteromalidae) at constant temperatures in China. J Med Entomol 56(2):368–377

Recent Developments in the Study of Forensic Flies

5

Abstract

Since its early history of study dating back to the thirteenth century in China, with the development of different techniques, the study of forensic flies has developed. There are several domains in which forensic entomology finds importance. In this chapter, we highlight the different areas of research focused on fly age and life cycle and cuticular hydrocarbons (CHCs) and other biochemical methods in the taxonomy and study of forensic flies. DNA based methods finds importance in the study of forensic flies. We also discuss the importance of artificial intelligence in the study of forensic flies.

Keywords

CHC · DNA · PCR · Molecular taxonomy · Artificial intelligence · Forensic entomology

Abbreviations

A. grahami	*Aldrichina grahami*
A. rufifacies	*Achoetandrus rufifacies*
ANN	Artificial neural networks
ATR-FTIR	Attenuated total reflection-Fourier transform infrared spectroscopy
B	*Boettcherisca peregrina*
B. germanica	*Blattella germanica*
C. albiceps	*Chrysomya albiceps*
C. bezziana	*Chrysomya bezziana*
C. chani	*Chrysomya chani*
C. idioidea	*Chloroprocta idioidea*

S. Ghosh, D. Banerjee, *Biology of Forensically Important Invertebrates*,
https://doi.org/10.1007/978-981-99-5026-3_5

C. macellaria	*Cochliomyia macellaria*
C. marginalis	*Chrysomya marginalis*
C. megacephala	*Chrysomya megacephala*
C. nigripes	*Chrysomya nigripes*
C. pinguis	*Chrysomya pinguis*
C. putoria	*Chrysomya putoria*
C. rufifacies	*Chrysomya rufifacies*
C. thanomthini	*Chrysomya thanomthini*
C. varipes	*Chrysomya varipes*
C. vicina	*Calliphora vicina*
C. villeneuvi	*Chrysomya villeneuvi*
C. vomitoria	*Calliphora vomitoria*
CHC	Cuticular hydrocarbons
F. canicularis	*Fannia canicularis*
GA-LDA	Genetic algorithm-linear discriminant analysis
GC-MS	Gas chromatography coupled to mass spectrometry
H. dentipes	*Hydrotaea dentipes*
H. ignava	*Hydrotaea ignava*
H. infumata	*Hypopygiopsis infumata*
H. ligurriens	*Hemipyrellia ligurriens*
H. melanura	*Helicophagella melanura*
H. pulchra	*Hemipyrellia pulchra*
H. similis	*Hydrotaea similis*
H. tumrasvini	*Hypopygiopsis tumrasvini*
HC	Hydrocarbon
L. ampullacea	*Lucilia ampullacea*
L. bazini	*Lucilia bazini*
L. caesar	*Lucilia caesar*
L. cuprina	*Lucilia cuprina*
L. eximia	*Lucilia eximia*
L. illustris	*Lucilia illustris*
L. papuensis	*Lucilia papuensis*
L. porphyrina	*Lucilia porphyrina*
L. retroversa	*Lucilia retroversa*
L. rica	*Lucilia rica*
L. sericata	*Lucilia sericata*
L. silvarum	*Lucilia silvarum*
L. sinensis	*Lucilia sinensis*
M. autumnalis	*Musca autumnalis*
M. domestica	*Musca domestica*
NIRS	Near-infrared spectroscopy
P. (Pattonella) intermutans	*Peckia (Pattonella) intermutans*
P. americana	*Periplaneta americana*
P. crassipalpis	*Parasarcophaga crassipalpis*
P. regina	*Phormia regina*

P. sericata	*Phaenicia sericata*
P. terraenovae	*Protophormia terraenovae*
P. terraenovae	*Protophormia terraenovae*
P. vulgaris	*Parapiophila vulgaris*
PCA	Principal component analysis
PMImin	Minimum postmortem interval
S. (Liopygia) ruficornis	*Sarcophaga (Liopygia) ruficornis*
S. africa	*Sarcophaga africa*
S. agnata	*Sarcophaga agnata*
S. albiceps	*Sarcophaga albiceps*
S. argyrostoma	Sarcophaga *argyrostoma*
S. calcitrans	*Stomoxys calcitrans*
S. carnaria	Sarcophaga *carnaria*
S. crassipalpis	Sarcophaga *crassipalpis*
S. dux	*Sarcophaga dux*
S. lambens	*Sarcodexia lambens*
S. Meigen	*Sarcophaga Meigen*
S. melanura	*Sarcophaga melanura*
S. nudiseta	*Synthesiomyia nudiseta*
S. peregrina	*Sarcophaga peregrina*
S. pumila	Sarcophaga *pumila*
S. subvicina	Sarcophaga *subvicina*
S. teretirostris	*Sarcophaga teretirostris*
S. tibialis	*Sarcophaga tibialis*
S. vagans	*Sarcophaga vagans*
S. variegata	*Sarcophaga variegata*
SPA-LDA	Successive projection algorithm-linear discriminant analysis

5.1 Introduction

Age determination of necrophagous fly of forensic importance across life cycle forms the basis of determination of the postmortem interval (PMI) and determination of the time when death has occurred (Sharma and Singh 2015).

Flies under the Order Diptera and Family Calliphoridae constitute a major family of forensically important insects and play a major role in estimation of minimum postmortem interval (PMImin) that aid in investigations of an act of crime and its detection.

In order to calculate PMImin, identification of the larval species present in the dead body finds importance. Identification of species involves use of morphological keys as the conventional method whereby different insects and life cycle stages can be identified but suffers from challenges of requiring long time-consuming analysis,

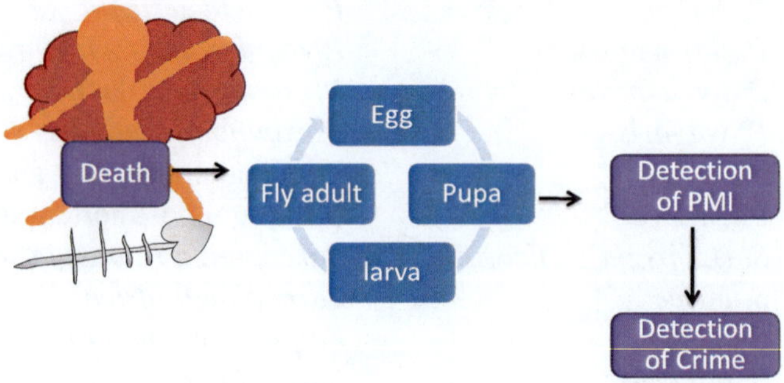

Fig. 5.1 Forensic entomology and detection of crime

being subjective, requiring trained manpower, and difficulty in identification of some stages of larval and pupal form within species by only morphological evidences (Fig. 5.1).

Thus, rapid identification of the immature larval stages finds importance in reducing time of identification of insects' stages that finds importance in criminal investigations. The empty puparia, left after metamorphosis of the immature to the adult fly, can act as important evidence at the site of crime and particularly finds importance in investigations of highly decomposed corpses. Thus, design of appropriate fast and reliable methods to determine the withering time of these puparial cases finds importance in detection of PMI. However, the estimation of PMI from insect evidences suffers from limitations and challenges (Amendt et al. 2021, 2011; Matuszewski 2021; Magni et al. 2012).

Several factors of the environment are known to regulate the growth of the flies on carcass. A study of colonization by using meat as bait by Victorian forensic Diptera species revealed nutritional content of meat as an important factor regulating the colonization of Diptera on carcass as compared to tested factors like larval digestive enzymes, excreta, and expression of cuticular hydrocarbons (CHC, George et al. 2012).

5.2 Forensic Entomology and Applications

Forensic entomology and its earliest studies date back to the thirteenth century in China where insects were used to detect crime. A farmer was killed in a rice field with a sickle but there was no evidence. When other suspected farmers were told to submit their respective sickles, one sickle among all suspects attracted numerous blow flies, indicated traces of blood, not visible by the naked eye, on the blade, highlighting the importance of entomological evidence, in detecting criminals (Sharma and Singh 2015; McKnight 1981).

Later on, several applications of forensic entomology were found encompassing any situation involving interaction between insects, arthropods, and law. The utilities have been categorized as urban entomology, stored crop/product entomology, and medicolegal forensic entomology (Sharma and Singh 2015; Hall 1990; Harvey 2006).

Table 5.1 Forensic entomology and applications[a]

Forensic entomology	Urban forensic entomology
	Stored product forensic entomology
	Medicolegal forensic entomology

[a]Sharma and Singh (2015), Hall (2001)

Urban entomology involves interaction of insects with anthropogenic structures and human society like termite/cockroach/insect infestation of buildings (Hall 1990), breeding of flies in livestock, etc. (Sharma and Singh 2015; Hall 2001).

The stored crop product forensic entomology involves the study of insect infestation in stored commodities/food/crops, invasion by insect pest on crops and domestic household products (Sharma and Singh 2015; Hall 2001) (Table 5.1).

5.3 Fly Age, Life Cycle, and Cuticular Hydrocarbons (CHCs)

Cuticular hydrocarbons (CHCs), located externally in insects, reveal species-specificity and are known to play a role in regulating physiology and ecological functions and confer protection from drying or desiccation and known to function as semiochemicals or organic compounds that relay specific chemical signals or messages that lead to modification of insect social behavior or physiology. In termites CHC is reported to vary within species, populations, castes, or seasons in quality and/or quantity (Gordon et al. 2020). CHCs in insects are largely analyzed by using gas chromatography coupled to mass spectrometry (GC-MS) and find application as tools in taxonomic differentiation of insect species and determination of insect sex and age of adult and immature forms.

GC-MS has been used in identification and quantification analysis of the CHCs of females and males of *Chrysomya putoria* (*C. putoria*, Wiedemann, 1818) under Order Diptera, family Calliphoridae of age ranging from 1 to 5 days old.

Although such study revealed expression of same major compounds in either sexes, expression of CHC varied with age (Braga et al. 2016).

CHC in insects revealing property of contact pheromones has been studied in insects of forensic importance, including blow flies under Order Diptera, family Calliphoridae, and is known to undergo rapid alteration with age. This finds importance in regulating their sexual behavior. Extracted CHC from necrophagous flies of family Calliphoridae like *Chrysomya rufifacies (C. rufifacies)*, commonly called as hairy maggot blow fly, a fly of forensic importance when analyzed by GC-MS approaches, revealed alteration of HCs with larvae age, being an important marker in insect development and enabling determination of age of the immature stages of specimens. This finds importance in estimating the time of colonization and inferring at the PMI (Sharma et al. 2021). CHC has been used to determine fly larvae age, and its changes have been observed in *C. rufifacies (Macquart)* larva using GC-MS. Most of the HCs identified indicate alkanes in nature with the carbon (C) chain length ranging from 21 to 31 Cs, with six different kinds of alkenes.

The expression of larval hydrocarbon (HC) correlated with the age of the larvae. Simple peak ratios of n-C29 and another eight selected peaks that increased with age indicated CHC as indicator for determination of age of *C. rufifacies* larvae at post-feeding stage. This finds particular importance as it is otherwise difficult to differentiate the stages only by means of morphological evidences (Zhu et al. 2006). Stable structures of CHC and their presence throughout blow fly life cycle starting from the egg to adult form enabled estimation of adult blow fly age. The CHC profiles of *Cochliomyia macellaria (C. macellaria, Fabricius)*, a native blow fly species, and *C rufifacies (Macquart)*, an invasive species of North America when studied under GC-MS approaches, revealed expression of 37 and 35 diverse HC compounds expressed by *C. macellaria* and *C. rufifacies*, respectively, of which 24 and 23 n-alkene carbon compounds and methyl-branched alkane HC macromolecules were identified from *C. macellaria* and *C. rufifacies*, respectively. Expression of ten compounds was observed in both species (Pechal et al. 2014) which revealed overlapping expression.

Chrysomya albiceps (*C. albiceps*, Wiedemann 1819), a member of family Calliphoridae under Order Diptera, another forensically important fly, reveals influences of abiotic, environmental, and ecological factors including nutrition, temperature of the environment, humidity, and variations due to geography affecting larval development and survival rates. Studies revealed body size and CHC as indicators in estimation of life age of different larval instars of *C. albiceps*. GC-MS studies revealed difference in CHC among different larval instar forms being reared at the same temperature and between the same larval instar being reared at different temperatures (Alotaibi et al. 2021). CHC of *Lucilia sericata (L sericata)*, *C. vicina*, and *Calliphora vomitoria (C vomitoria)*, under Order Diptera, family Calliphoridae, extracted from the cuticles of adult flies and analyzed using GC-MS revealed alterations in hydrocarbon expression between the immature flies and mature adult flies (Moore et al. 2017).

CHC extracted from blow fly larva of *Calliphora vicina (C. vicina)* and *C. vomitoria* analyzed by GC-MS via statistical methods of principal component analysis (PCA) and computational approaches of artificial neural networks (ANN) revealed alterations in their chemical profiles that can infer the larval age ranging from younger larvae to larvae at post-feeding stage (Moore et al. 2016, 2017a).

CHC n-pentacosane or n-C25 expressed in the CHC, extracted from the legs of the adult flies of L *sericata* and *C vicina* identified by GC-MS analysis, was found to be an important tool in age estimation of blow flies and known to play a role in forensic entomology (Bernhardt et al. 2017a, b). CHC of *L sericata, C. vicina*, and *C. vomitoria*, under Order Diptera, family Calliphoridae, analyzed by GC-MS analysis revealed differences between the immature flies and mature adult flies (Moore et al. 2017a). CHC studied by GC-MS of *C. vicina* and *L. sericata* puparial cases revealed age of flies and finds importance as indicators of PMI (Moore et al. 2013) and CHC hydrocarbons gained importance in identifying Sarcophagidae in the studies of Moore et al. (2021).

In another study it has been reported that CHC of forensically important Dipterans under family Calliphoridae, studied by GC analysis including

C. vomitoria (Linné), C. vicina (Robineau-Desvoidy), and *Protophormia terraenovae (P terraenovae, Robineau-Desvoidy)*, from eggs to adults (8-day-old), revealed major expression of odd linear chain of alkanes. Short-chain carbon (C) compounds expressed in fly larvae and post-feeding larvae were observed to be developed to long-chain carbon (C) compounds in pupae and adult forms, thus highlighting the importance of CHC in cuticular waxes in Calliphoridae in determination of PMI with roles in forensic sciences (Roux et al. 2008) .Epicuticular HC content studied from 3 to 120 h post-emergence of larva in flies of both sexes of *C vomitoria* revealed expression of n-Alkanes and mono- and dimethyl alkanes at a temperature range of 20–31 °C forming the predominant form of HC. Alkenes were only detected in the females, while males expressed fewer HC, revealing di- and monomethylalkanes with 23 C, 24 C, and 26 C atoms (Trabalon et al. 1992).

Volatile HC from larvae and pupae of *C vicina* Robineau-Desvoidy analyzed by GC-MS revealed expression of both un-branched and branched HC, alcohols, acids, and esters, varying in their composition or quality and quantity, with that of the age of the developmental stages of larva or pupa, with a probable role in PMI estimation (Frederickx et al. 2012). *C vicina* adults revealed the increased pteridine, metabolites accumulating in the eyes during the process of aging, with accumulated degree-days (ADD), and revealed expression differences in between male and female flies (Bernhardt et al. 2017b). HC extracted from the cuticle of the stage till the pupae develops for *L. sericata* when analyzed by GC-MS revealed study of the age of larvae throughout life cycle, highlighting the importance of GC-MS technique with a vital role in study of PMI (Moore et al. 2013).

Temperature at which larvae is reared revealed to affect the expression of CHC in insects of forensic importance. GC and GC-MS analysis of CH of larva of Aldrichina *grahami* (*A grahami*, Aldrich) under Order Diptera and family Calliphoridae when reared at 24 °C revealed the age-dependent, quantitative changes in HCs, revealing presence of alkanes with chain length less than or equal to C25 that reduced with larval development, but alkanes of chain length more than C25 increased in expression with increase in age of the developing stage. Expression of pentacosene a, n-C25, n-C27, and n-C31 was found to be of utmost significance, while the peak ratio n-C29 divided by each chromatographic peak revealed correlation with larvae age (Xu et al. 2014).

In *Chrysomya varipes* (*C. varipes*), commonly called as small hairy maggot blow fly, CHCs revealed age- and sex-specific alterations in expression which has been associated with maturation of adult blow fly occurring more slowly in females (Butterworth et al. 2020a,b) as compared to that of males.

CHC profiles of house flies, *Musca domestica L. (Musca domestica),* under Order Diptera, family Muscidae, collected in South end of California, Georgia, and Minnesota studied for gonotrophic ages of females revealed expression of (Z)-9-tricosene (muscalure, Butler et al. 2009). Biophysical studies on the cuticular lipids of housefly, *M domestica*, revealed decrease in melting temperatures (Tm) in lipids extracted from cuticle of female houseflies from 39.3 °C to 35.3 °C on attaining sexual maturity and production of sex pheromone.

Lipids from males, however, did not reveal changes with age. Cuticular lipids revealed expression of saturated, unsaturated, and methyl-branched forms of hydrocarbons (HCs, Gibbs et al. 1995). Adult female *M domestica L.* in California, Georgia, and Minnesota revealed expression of (Z)-9-tricosene (muscalure) in abundance among the dominant eight HC expressed (Butler et al. 2013).

Morphology-based identification of pupae and adult stage of *Sarcophaga peregrina* (*S. peregrina* Robineau-Desvoidy, 1830), belonging to Sarcophagidae family of flesh fly species found globally, by its morphology, remains challenging. CHC of *S. peregrina* from larval to adult stage analyzed by GC-MS approaches revealed expression of 63 compounds with carbon (C) of chain length ranging from 8 to 36, comprising of n-alkanes, branched alkanes, alkenes, and unknowns. With development, branched alkane forms revealed a dominant and variable expression of high-molecular weight (hmw) HC, including expression of 2-methyl C19, Di-methyl C21, docosane (C22), and tricosane (C23, Zhang et al. 2022) indicating the importance of CHC in identification of the life cycle stage.

5.4 CHC and Taxonomy of Forensic Flies

The expression and quantity of CHCs in insects can be species-specific and known to reveal sexual dimorphism within species.

CHC of adult male flies and female flies of flesh flies including *Sarcophaga Meigen* (*S Meigen* Sarcophagidae), *Sarcophaga africa* (*S Africa* Wiedemann), S. *agnata Rondani, S. argyrostoma Robineau-Desvoidy, S. carnaria (Linnaeus), S. crassipalpis Macquart, S. melanura Meigen, S. pumila Meigen, S. teretirostris Pandellé, S. subvicina Rohdendorf, S. vagans Meigen*, and *S. variegata* (Scopoli) from Natural History Museum, London, using GC-MS enabled reliable identification of a most specimens, even from very old collections revealing sex-specific differences within species (Moore et al. 2021).

GC-MS finds importance in identification of CHCs from four species of *Sarcophagidae*, Peckia (Peckia) *chrysostoma*, *Peckia (Pattonella) intermutans*, *Sarcophaga (Liopygia) ruficornis*, and *Sarcodexia lambens*, revealing expression of mixture of n-alkanes, mono-methyl alkanes, and dimethyl alkanes with linear carbon (C) chain lengths varying from 23 to 33C atoms with expression of two alkenes only from all four species. *S. lambens* revealed linear chain lengths varying from C23 to C33, while the other three species revealed HC of linear chain lengths 24 to 31C atoms. All four species revealed expression of n-Heptacosane with molecular formula $C_{27}H_{56}$, (straight-chain with 27 carbon alkane), n-nonacosane with molecular formula $C_{29}H_{60}$ (straight-chain alkane of 29 c atoms), and 3-methylnonacosane with formula $C_{30}H_{62}$, n-triacontane with molecular formula $C_{30}H_{62}$, and n-hentriacontane with molecular formula $C_{31}H_{64}$ (long-chain alkane). This study highlighted the importance of HC profiles in taxonomic differentiation of insect species (Moore et al. 2021) (Fig. 5.2).

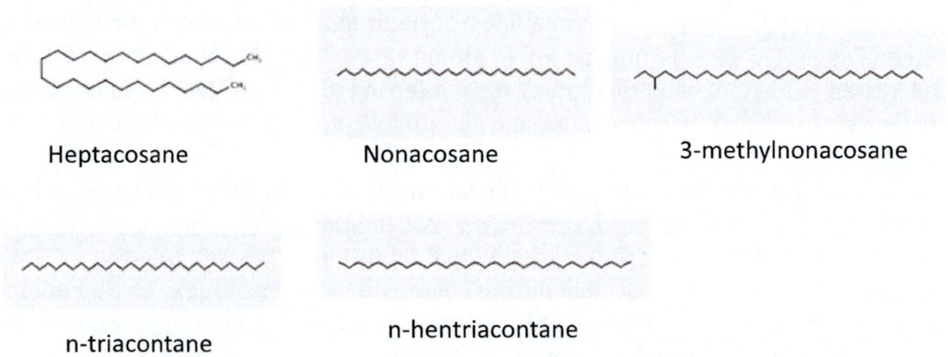

Fig. 5.2 Heptacosane, https://www.ebi.ac.uk/chebi/searchId.do?chebiId=CHEBI:32941), nonacosane (https://pubchem.ncbi.nlm.nih.gov/compound/Nonacosane), 3-methylnonacosane, (https://pubchem.ncbi.nlm.nih.gov/compound/3-Methylnonacosane), n-triacontane (https://pubchem.ncbi.nlm.nih.gov/compound/Triacontane#section=Canonical-SMILES) and n-hentriacontane (https://pubchem.ncbi.nlm.nih.gov/compound/Hentriacontane)

CHC in genus *Chrysomya* reveal diversity in terms of quality and quantity in expression of contact pheromones like CHCs with age and contribute to sex development (Butterworth et al. 2020a, b).

In a study on forensic flies, GC-MS was used to analyze the CHC extracted from the pupal exuviae of six necrophagous flies, including *A. grahami (Aldrich)*, *Achoetandrus rufifacies* (Macquart), *Boettcherisca peregrina (Robineau-Desvoidy)*, *Chrysomya megacephala* (F.), *L. sericata (Meigen),* and *Parasarcophaga crassipalpis (Macquart)*, revealing importance of CHC as chemotaxonomic tools in study of insects of forensic importance (Ye et al. 2007).

Adults of *C megacephala* (F. 1794) isolated from two different locations in Brazil revealed variation in CHC profiles between the two populations, developmental stages, thus highlighting its importance as a technique for identification of species and its stages, together with geographical variability (Paula et al. 2017).

CHC together with fatty acids of eggs of flies of the family Calliphoridae can be used to distinguish species (Lunas et al. 2019).CHC analysis on first instar larvae of *C vicina*, *L sericata,* and *C vomitoria* using GC-MS allowed accurate identification when it can be challenging to study through morphological keys (Moore et al. 2014).

The CHC of empty puparia from flies of forensic importance including *C vicina*, *C. albiceps*, *Lucilia caesar*, Protophormia terraenovae, *Lucilia silvarum*, *Phormia regina, L sericata,* and the flesh fly *Sarcophaga caerulescens* reveal differences in the expression of CHC enabling identification of all species (Moore et al. 2022).

Calliphoridae puparial cases can establish the age inferring at the PMI. *C vicina* and *L sericata*, HCs analysed by GC-MS, revealed as important tools in determining age differences within the empty cases (Moore et al. 2017b).

GC-MS of CHC of the cases of pupae of *C. megacephala* revealed expression of mixture of n-alkanes, methyl-branched alkanes, and dimethyl-branched alkanes, wherein the C atom chain length ranged from C21 to C35 atoms. Expression of n-C22, n-C24, and n-C26 revealed an increase with withering time. While n-C26 expression revealed an increase in a linear fashion in abundance with the withering time. For n-C26 or less number of C atoms in the HC, the abundance of CHC decreased with the withering time, thus inferring that CHC could indicate the withering time in *C. megacephala* and may play role in PMI determination (Zhu et al. 2007).

A study on *Phormia regina (P. regina)* from different geographical locations including Tucannon River, Lyle Grove, Washington, and Rensselaer, Indiana, revealed that CHC extracted from individual adult blow flies subjected to GC-MS revealed expression of 22 HC that differed based on location and sex. These findings highlighted importance of geography of a location of corpse and impact on ecology of population of forensic fly (Byrne et al. 1995).

5.5 Other Biochemical Methods in the Study of Forensic Flies

The insect species and its age, observed at the scene of crime, can be used in estimating the PMI. Conventional methods of species and developmental stage identification suffer from challenges of being subjective, costly, and time-consuming requiring the expertise or molecular DNA analysis. Vibrational spectroscopy offers quick, cheap, and less subjective technique, in a study of entomological evidence in forensic process. Attenuated total reflection-Fourier transform infrared spectroscopy (ATR-FTIR) has been used to differentiate species of larvae of forensically important species isolated from crime scenes (Pickering et al. 2015).Near-infrared spectroscopy (NIRS), a rapid, cheap, and nondestructive method of metabolomic profiling, is used for discriminating closely related species of same genus, and ATR-FTIR spectroscopy is used to get an infrared spectrum together with variable selection techniques like genetic algorithm-linear discriminant analysis (GA-LDA) and successive projection algorithm-linear discriminant analysis (SPA-LDA) in identification of six species of flesh flies from neotropical regions from 160 NIRS and FT-IR specimens. The genera *Oxysarcodexia*, *Peckia*, and *Ravinia* revealed differences, and this tool found importance as a fast and nondestructive method in identification of forensic flies (Barbosa et al. 2018).

To circumvent the limitations of DNA-based studies of forensic entomology taxonomy, newer strategies based on immunological antigens are being exploited to find out differences between different species of blow flies. Differences in expression in surface antigens within first instar larva of four forensically important blow flies have revealed expression of immunoreactive proteins which may be used in the development of species-specific immuno-diagnostic tests for identification of forensically important flies (McDonagh et al. 2009).

5.6 DNA-Based Methods and Its Application in Forensic Entomology

One of the most common molecular genotyping methods used in taxonomy is DNA-based identification of specimen. Recently RNA- and DNA-based methods are being used for molecular analysis of insect gut contents and characterization of forensically important insect species (Wells and Stevens 2008).

Molecular tools find particular importance in identification of cryptic species or specimens that are poorly conserved and their identification like females of Sarcophagidae species for which morphological identification is difficult. In a study molecular identification has been reported by sequencing the COX-I mitochondrial gene and analyzing the sequences which could lead to identification of forensic flies including *Sarcophaga argyrostoma* and *Sarcophaga tibialis*, and this method finds importance in identifying Sarcophagidae for forensic investigations and use for estimation of minPMI (Fuentes-López et al. 2021).

Using mitochondrial cytochrome *c* oxidase subunit I (COI) and COII sequences, 16 species of blow flies in Thailand including *Chrysomya bezziana*, *Chrysomya chani*, *C. megacephala*, *Chrysomya nigripes*, *Chrysomya pinguis*, *C. rufifacies*, *Chrysomya thanomthini*, *Chrysomya villeneuvi*, *Lucilia cuprina*, *Lucilia papuensis*, *Lucilia porphyrina*, *Lucilia sinensis*, *Hemipyrellia ligurriens*, *Hemipyrellia pulchra*, *Hypopygiopsis infumata*, and *Hypopygiopsis tumrasvini* could be detected (Sontigun et al. 2018).

DNA barcoding, using COI and ITS2, has enabled fast and precise identification of Calliphoridae in a study on blow flies from the Caribbean region with 24 distinct clades with substantial geographical variation for *Lucilia eximia, Lucilia retroversa, Lucilia rica*, and *Chloroprocta idioidea* probably indicating occurrence of several cryptic species of blow flies (Yusseff-Vanegas and Agnarsson 2017).

Specimens of Muscidae subjected to different analysis of semiautomated identification, wing measurements, and DNA-based approaches revealed identification of flies under genus including *Azelia*, *Mydaea*, *Eudasyphora*, *Graphomya*, *Musca*, *Stomoxys*, *Muscina*, *Neomyia*, *Polietes*, *Hydrotaea*, and *Thricops* (Grzywacz et al. 2017). Mitochondrial DNA sequencing data and phylogenetic analysis were used to identify blow fly subfamily Chrysomyinae collected from a dead body in Canada or the USA (Wells and Sperling 2001). Wing shape variation and DNA-based methods find importance in identifying both Calliphoridae and Sarcophagidae (Khang et al. 2021).

Forensic flies collected from specimens from corpses, in a study reported from Portugal, studied by the DNA sequencing of mitochondrial COX-1 gene could aid in identification of Calliphoridae and Sarcophagidae (Cainé et al. 2009).

DNA-based COI sequence has been applied as a suitable DNA marker in identification of Calliphoridae (Meng et al. 2017). In a DNA-based taxonomic study of flow flies, COI and COII sequences have been used successfully for identification of Sarcophagidae in a study from China (Zhang et al. 2015) and COI for Sarcophagidae from India (Sharma et al. 2015) and Australia (Meiklejohn et al. 2011). DNA-based identification of COI gene from Calliphoridae in Malaysia

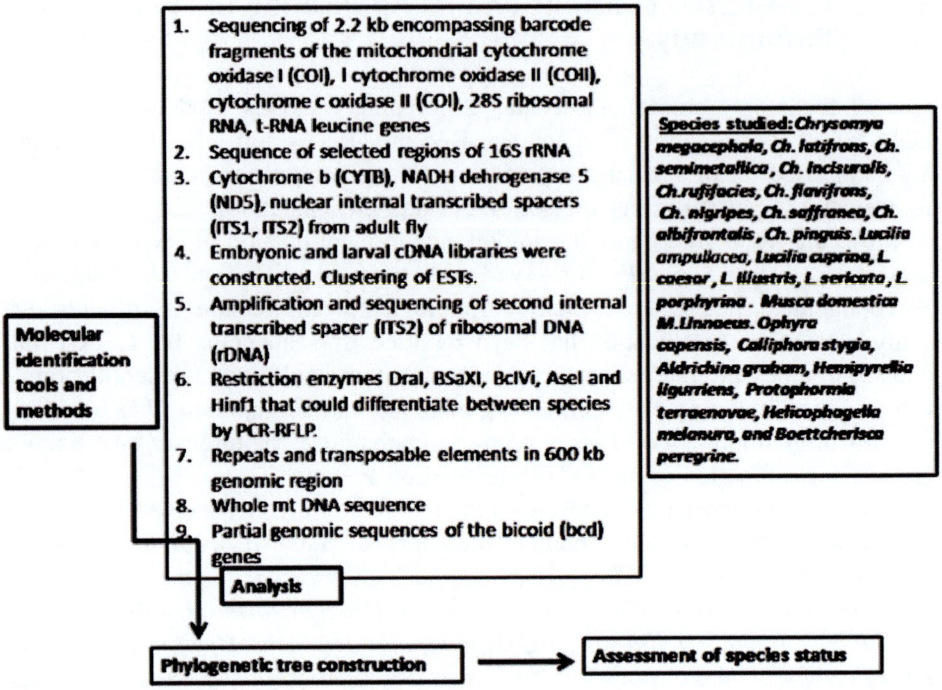

Fig. 5.3 Molecular identification of forensic fly and its immature stages. Image adapted from Ghosh et al. (2018)

enabled identification of two species, *C megacephala* and *C rufifacies*, from human dead bodies (Kavitha et al. 2013a, b). COI gene sequence has been used as a tool for successful identification of three Sarcophagidae from Rio de Janeiro including *Oxysarcodexia fluminensis, Peckia chrysostoma*, and *Peckia intermutans* (Napoleão et al. 2016). Sonication has been used as a method in DNA extraction for fast, reliable, and efficient nondestructive DNA extraction from carrion flies (Stamper et al. 2017) (Fig. 5.3).

The molecular identification methods are based on DNA sequencing or PCR-RFLP analysis of different partial or total genetic sequences. The differences in the sequences find importance in phylogenetic tree analysis to rule out intra- and interspecific variations (Balme et al. 2012; Chin et al. 2009; Lang et al. 2006; Lee et al. 2011; Gilarriortua and Salona 2013; Ghosh et al. 2018; Griffiths et al. 2009; Harvey et al. 2008; Kavitha et al. 2012; Negre and Simpson 2013; Nelson et al. 2008; Park et al. 2013; Preativatanyou et al. 2010; Sonet et al. 2012; Yang et al. 2010; Zaidi et al. 2011).

PCR-based methods on insect larvae from corpse for the COI, COII, and tRNA-leucine genes have been used for estimating PMI and have been applied for rapid identification of *Phormia regina, Phaenicia sericata*, and *Lucilia illustris* (Sperling et al. 1994).

DNA-based methods, age estimation by morphological analyses, and differential gene expression (DGE) studies have enabled identification of *C. vicina* in a study from 4- and 7-day-old pupae (Brown et al. 2012). DNA identification of blow fly species from mitochondrial DNA from whole blow fly and legs, egg, first instar, second instar, third instar larva, puparium and empty puparium for every life cycle stage has been used to identify *C. megacephala (Fabricius)* and *C. rufifacies (Macquart)* from Malaysia (Kavitha et al. 2013a, b). Sequencing and molecular techniques using COI gene have been used for identification of Chrysomyinae including *C. albiceps (Wiedemann), C. megacephala (F.),* and *Chrysomya marginalis (Wiedemann)]* (Diptera: Calliphoridae) from Egypt revealing divergence (Salem et al. 2015).

Common molecular biology and DNA-based species determination methods suffer from the failure of distinguishing *Lucilia* Robineau-Desvoidy as *Lucilia illustris* and *L. caesar*, reported to occur in Europe. *L. illustris* specimens collected from North America are difficult to diagnose, due to shared mitochondrial and nuclear gene sequences. Thus, amplified fragment length polymorphism (AFLP) is used as a better tool (Picard et al. 2017). DNA-based methods based on COI gene have been used to identify six parasitic invasive phorid fly species including *Conicera tibialis*, *Megaselia scalaris*, *Megaselia abdita*, *Megaselia giraudii, Megaselia rufipes*, and *Puliciphora borinquenensis* to identify several species of scuttle flies (Boehme et al. 2010).

COI-I and COI-II genes have been used for discriminating Sarcophagidae from China and Egypt including *Sarcophaga albiceps* (Meigen), *Sarcophaga argyrostoma (Robineau-Desvoidy), Wohlfahrtia nuba (Wiedemann, Aly and Wen* 2013*)*, and *Sarcophaga dux (Thomson, Aly* et al. 2013*)*. COI gene has been used to identify Calliphoridae from China including Chrysomya nigripes, *C. megacephala, C. albiceps, C. rufifacies,* and *A. grahami*; *Lucilia bazini, Lucilia caesar, Lucilia cuprina, L sericata,* and *L porphyrina*; members of *Muscidae* including *Musca autumnalis, M. domestica,* lesser house fly *Fannia canicularis,* and stable fly *Stomoxys calcitrans*; and members of Sarcophagidae including *Sarcophaga albiceps, Sarcophaga dux, Helicophagella melanura*, and in Egypt including members of Calliphoridae such as *C. megacephala, C. albiceps, L. sericata*; members of *Muscidae* including *M. domestica, F. canicularis, S. calcitrans, Synthesiomyia nudiseta*; and members of Sarcophagidae including flesh fly *Sarcophaga argyrostoma* (Aly and Wen 2013).

COI gene barcoding has been used to identify species from Germany including *C. vicina, C. vomitoria, Lucilia ampullacea, Lucilia caesar, L. illustris, Lucilia silvarum, Phormia regina, Protophormia terraenovae, Piophilidae: Parapiophila vulgaris, Muscidae: Hydrotaea dentipes, Hydrotaea ignava,* and *Hydrotaea similis* (Boehme et al. 2012). COII and 16S ribosomal RNA (16S rRNA) genes have been used to identify Sarcophagidae (Guo et al. 2012) in a study reported from China (Fig. 5.4).

5.7 Artificial Intelligence and Study of Forensic Entomology

Digital forensics is a newly developing branch of science that involves large-scale processing, computing, and analysis of huge and complicated datasets. Forensic investigation requires proper analysis of information gathered from many sources and their integration toward a proper conclusion. Artificial intelligence (AI) and

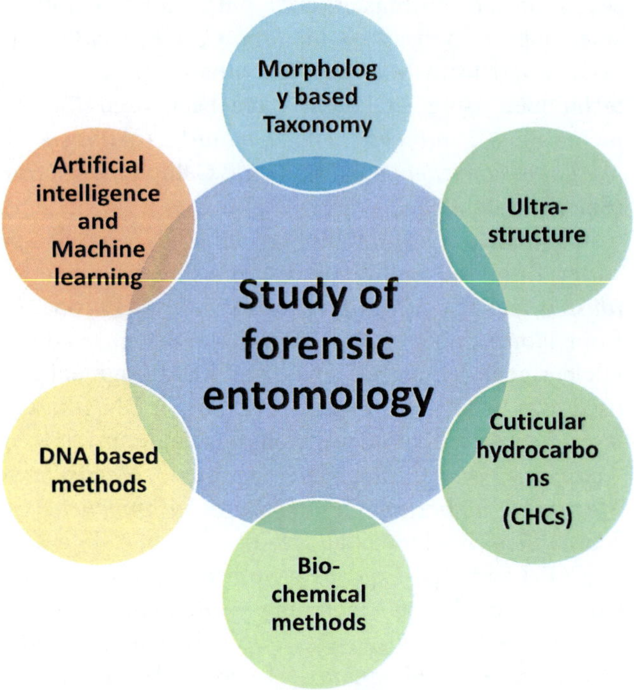

Fig. 5.4 Different approaches used in the study of forensic entomology. Image adapted from open access article (Matuszewski 2021)

machine learning (ML) have a major application in biological sciences. Recently, a lot of progress has been in the domain of artificial intelligence (AI) and its application in forensic sciences, leading to the application in digital forensics. As digital forensics is focused on computing and analyzing large volumes of data in its complexity within short period of time and AI is a promising tool in circumventing some of the limitation in digital forensics (Mohsin and kamshad 2021), artificial intelligence finds application in natural language processing, speech recognition, and machine vision and function by analyzing fast and high-throughput labeled training data, image recognition, correlations, and pattern recognition and then using these patterns to forecast future states (Ghosh and Dasgupta 2022).

Sensor-based monitoring of insects, image-based identification of insects, development of databases, generation of training data, and applications of deep learning (DL) tools are the latest development of machine learning and artificial intelligence in the field of ecological sciences and have been employed to address questions that remain hitherto unanswered in entomology (Høye et al. 2021).

A study in Mexico has revealed the application of computer vision and AI techniques of feature extraction, together with image processing and application of artificial neural network (ANN) algorithm that has enabled identification of cockroach pronotum from the images, from their color, shape, and texture leading to identification of three species of cockroaches including *Blattella germanica*, *Periplaneta Americana*, and *Arenivaga* sp. (Luquin et al. 2017).

Circumventing the limitations of conventional methods in the identification of species of fly, convolutional neural networks (CNNs) and processing of images of posterior spiracles have been used to identify maggots of four species including *L cuprina*, *C megacephala*, *C rufifacies*, and *M domestica* (Apasrawirote et al. 2022).

Statistical approaches and Bayesian framework (BF) have been used to evaluate complex hypotheses from observation on different insect fly species, accounting for uncertain rate of development, temperature required, and pre-colonization conditions. This is a development in forensic entomology in evaluating forensic results postdevelopment of the chemical, biological, radiological, and nuclear (CBRN) incident (Andersson et al. 2013).

The predominant parameter that finds application in age estimation is the boost in larval size and length during process of growth that can be compared to species-specific reference data to provide a larval age estimate. Recently, statistical methods have found importance in deriving inferences from such data. In Europe, *Calliphora vicina* is a predominant forensic insect, but lack of data is observed with respect to insect developmental data. Local data on development of forensic fly *C. vicina* from Germany for the larval stages, grown at three constant temperatures of 15, 20, and 25 °C, was used to compare with date on development of larva from other places and analyzed by quantile regression analysis by the linear model of development. This way of analysis reveals difficulty in describing post-feeding stage with linear models and calls in further research. These studies further highlighted the important role of collection of data of growth rates for local blow fly populations and requirement of standardizing experimental designs and statistical analyses at this time point (Baqué et al. 2015).

5.8 Challenges in the Study of Forensic Entomology

There are different areas of challenges related to the study of forensic entomology including (a) death, (b) pre-appearance intervals, (c) presence intervals, (d) developmental intervals, (e) post-eclosion interval, (f) successional stages, and (g) collection of evidences (Fig. 5.5).

The difficulties and challenges faced by forensic entomologists have been largely reported from local areas. For instance, studies have shown the difficulties and challenges from China (Wang et al. 2021), South America, and Brazil (Gomes and Von Zuben 2006).

Research in forensic entomology is enriched by findings both from research and casework. Yet casework suffers from limitations of localized study, poor sample preservation, less or late availability of data to others far away from the case study, and differences from replicated, controlled research work (Hall 2021).

Forensic practitioners face numerous challenges relating data findings with law in its terminology and description of insect age, and its relevance to colonization time of flies on remains, and limitations to making of estimates. Based on nationwide differences in legal requirements, neither standard format for the entomological case reports exist, nor any essential, guidelines exist on essential, desirable or not

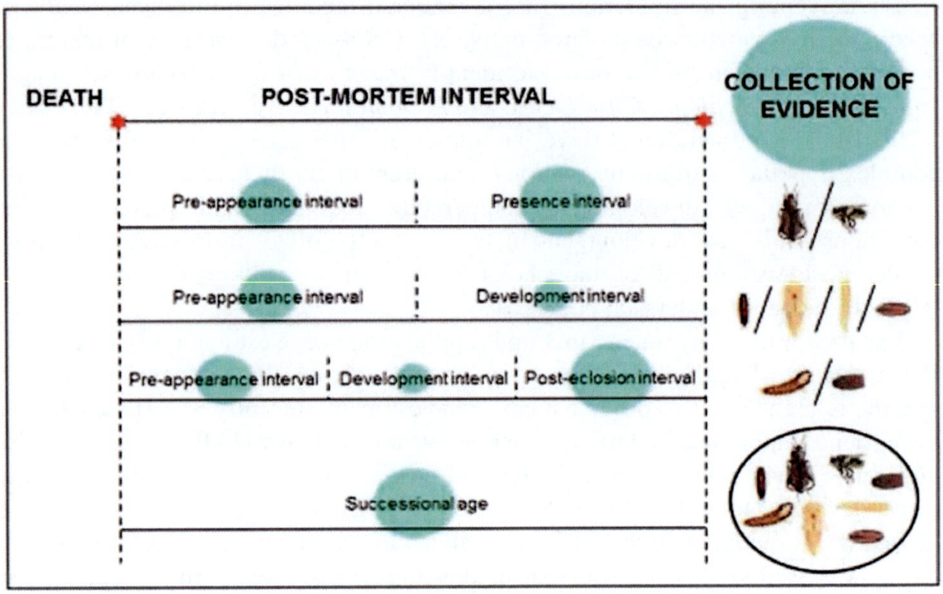

Fig. 5.5 Challenges in conventional forensic entomology. Image adapted from open-access article (Matuszewski 2021)

applicable information from a case report and global, standardized entomological case report does not exist. Thus, selection and setting of criteria for entomological evidence and their limitations remain major challenges in forensic medicolegal entomological sciences and need collaborative efforts by scientists, entomologists, and legal practitioners globally to finalize a guideline (Kotzé et al. 2021). The impact of poisons on insects is used for PMI estimation (Chophi et al. 2019).

Identification of forensically important insects is based on the use of dichotomous keys and morphological characteristics. Morphological identifications suffer from challenges of nonavailability of local keys and difficulty of use in identifying young and juvenile stages of forensic flies. For a damaged specimen, certain keys are non-usable for identification, and molecular identification suffers advantages in identification of forensically important insects, in different life stages or specimen completeness. Several methods including genetic distance and methods based on characters or reciprocal monophyly have been used in forensic identification studies. (Gemmellaro et al. 2019).

PMI estimations entail complicacy arising out of inter-individual variation in human decomposition and therefore reporting of instances where blow fly colonization fails to align with expected decomposition of soft tissue as blow fly larvae finds importance to produce estimation of time of colonization (TOC) to conclude a PMImin. Nonlinear nature of human decomposition contributes to further challenges to interpretation of blow fly evidence (Owings et al. 2022).

5.9 Discussions

With the progress of research, in this field, application of new techniques involving age-specific molecular profiles based on gene and protein expression near-infrared (IR) spectroscopy, used for qualitative and quantitative analyses, has been used in forensic entomology (Amendt et al. 2021).

Although forensic entomology involves collecting, identifying, and analyzing insect evidence, it still suffers from challenges of analysis by trained manpower for accurate and convincing results to convince a crime in a court (Amendt et al. 2011, 2021).

Matuszewski in an article in 2021 (Matuszewski 2021) has highlighted the different major challenges associated with estimation inaccuracy encompassing both major areas related to (i) collection and (ii) analysis of insect as evidence (Fig. 5.3). This paper also focused on the need for (i) better methods in collection of insect evidence, (ii) requirement of focused research on insect development and succession, (iii) the importance of reconstructing temperature conditions, and (iv) analysis of evidence followed by its relation to PMI estimation.

References

Alotaibi F, Alkuriji M, AlReshaidan S, Alajmi R, Metwally DM, Almutairi B, Alorf M, Haddadi R, Ahmed A (2021) Body size and cuticular hydrocarbons as larval age indicators in the forensic Blow Fly, Chrysomya albiceps (Diptera: Calliphoridae). J Med Entomol 58(3):1048–1055

Aly SM, Wen J (2013) Applicability of partial characterization of cytochrome oxidase I in identification of forensically important flies (Diptera) from China and Egypt. Parasitol Res 112(7):2667–2674

Aly SM, Wen J, Wang X (2013) Identification of forensically important Sarcophagidae (Diptera) based on partial mitochondrial cytochrome oxidase I and II genes. Am J Forensic Med Pathol 34(2):159–163

Amendt J, Richards CS, Campobasso CP, Zehner R, Hall MJ (2011) Forensic entomology: applications and limitations. Forensic Sci Med Pathol 7(4):379–392

Amendt J, Bugelli V, Bernhardt V (2021) Time flies-age grading of adult flies for the estimation of the post-mortem interval. Diagnostics (Basel) 11(2):152

Andersson MG, Sundström A, Lindström A (2013) Bayesian networks for evaluation of evidence from forensic entomology. Biosecur Bioterror 11(Suppl. 1):S64–S77

Apasrawirote D, Boonchai P, Muneesawang P et al (2022) Assessment of deep convolutional neural network models for species identification of forensically-important fly maggots based on images of posterior spiracles. Sci Rep 12:4753

Balme GR, Denning SS, Cammack JA, Watson DW (2012) Blow flies (Diptera: Calliphoridae) survive burial: evidence of ascending vertical dispersal. Foren Sci Int 216:e1–e4

Baqué M, Filmann N, Verhoff MA, Amendt J (2015) Establishment of developmental charts for the larvae of the blow fly Calliphora vicina using quantile regression. Foren Sci Int 248:1–9

Barbosa TM, de Lima LAS, Dos Santos MCD, Vasconcelos SD, Gama RA, Lima KMG (2018) A novel use of infra-red spectroscopy (NIRS and ATR-FTIR) coupled with variable selection algorithms for the identification of insect species (Diptera: Sarcophagidae) of medico-legal relevance. Acta Trop 185:1–12

Bernhardt V, Hannig L, Kinast R, Verhoff MA, Rothweiler F, Zehner R, Amendt J (2017a) Quantitative pteridine fluorescence analysis: a possible age-grading technique for the adult stages of the blow fly Calliphora vicina (Diptera: Calliphoridae). J Insect Physiol 98:356–359

Bernhardt V, Pogoda W, Verhoff MA, Toennes SW, Amendt J (2017b) Estimating the age of the adult stages of the blow flies Lucilia sericata and Calliphora vicina (Diptera: Calliphoridae) by means of the cuticular hydrocarbon n-pentacosane. Sci Justice 57(5):361–365

Boehme P, Amendt J, Disney RH, Zehner R (2010) Molecular identification of carrion-breeding scuttle flies (Diptera: Phoridae) using COI barcodes. Int J Legal Med 124(6):577–581

Boehme P, Amendt J, Zehner R (2012) The use of COI barcodes for molecular identification of forensically important fly species in Germany. Parasitol Res 110(6):2325–2332

Braga MV, Pinto ZT, Queiroz MM, Blomquist GJ (2016) Effect of age on cuticular hydrocarbon profiles in adult Chrysomya putoria (Diptera: Calliphoridae). Forensic Sci Int 259:e37–e47

Brown K, Thorne A, Harvey M (2012) Preservation of Calliphora vicina (Diptera: Calliphoridae) pupae for use in post-mortem interval estimation. Forensic Sci Int 223(1–3):176–183

Butler SM, Moon RD, Hinkle NC, Millar JG, McElfresh JS, Mullens BA (2009 Dec) Characterization of age and cuticular hydrocarbon variation in mating pairs of house fly, Musca domestica, collected in the field. Med Vet Entomol 23(4):426–442

Butler SM, Moon RD, Hinkle NC, Millar JG, McElfresh JS, Mullens BA (2013) Gonotrophic development and survival in field populations of Musca domestica (Diptera: Muscidae) at dairies in California, Minnesota, and Georgia, and the relationship of fly age to relative abundance of (Z)-9-tricosene (muscalure). J Med Entomol 50(4):748–757

Butterworth NJ, Drijfhout FP, Byrne PG, Keller PA, Wallman JF (2020a) Major Transitions in cuticular hydrocarbon expression coincide with sexual maturity in a Blowfly (Diptera: Calliphoridae). J Chem Ecol 46(7):610–618

Butterworth NJ, Wallman JF, Drijfhout FP, Johnston NP, Keller PA, Byrne PG (2020b) The evolution of sexually dimorphic cuticular hydrocarbons in blowflies (Diptera: Calliphoridae). J Evol Biol 33(10):1468–1486

Byrne AL, Camann MA, Cyr TL, Catts EP, Espelie KE (1995) Forensic implications of biochemical differences among geographic populations of the black blow fly, Phormia regina (Meigen). J Foren Sci 40(3):372–377

Cainé LM, Real FC, Saloña-Bordas MI, de Pancorbo MM, Lima G, Magalhães T, Pinheiro F (2009) DNA typing of Diptera collected from human corpses in Portugal. Foren Sci Int 184(1–3):e21–e23

Chin HC, Ahmad NW, Lim LH, Jeffery J, Omar B, Dhang CC, Weng LK, Sofian-Azirun M (2009) Research Note Predation on pupa of Chrysomya rufifacies (Marquart)(Diptera: Calliphoridae) by parasitoid, Exoristobia philippinensis Ashmead (Hymenoptera: Encyrtidae) and Ophyra spinigera larva (Diptera: Muscidae). Trop Biomed 26(3):369–372

Chophi R, Sharma S, Sharma S, Singh R (2019) Forensic entomotoxicology: current concepts, trends and challenges. J Forensic Legal Med 67:28–36

Frederickx C, Dekeirsschieter J, Brostaux Y, Wathelet JP, Verheggen FJ, Haubruge E (2012) Volatile organic compounds released by blowfly larvae and pupae: new perspectives in forensic entomology. Forensic Sci Int 219(1–3):215–220

Fuentes-López A, Pedreño Sala A, Romera E, Galián J (2021) DNA-based and taxonomic identification of forensically important Sarcophagidae (Diptera) in southeastern Spain. Sci Justice 61(2):150–159

Gemmellaro MD, Hamilton GC, Ware JL (2019) Review of molecular identification techniques for forensically important diptera. J Med Entomol 56(4):887–902

George KA, Archer MS, Toop T (2012) Effects of bait age, larval chemical cues and nutrient depletion on colonization by forensically important Calliphorid and Sarcophagid flies. Med Vet Entomol 26(2):188–193

Ghosh S, Dasgupta R (2022) Machine learning in biological sciences: updates and future prospects. Springer Nature. ISBN: 978–981–16-8881-2

Ghosh S, Ansar W, Banerjee D (2018) Diagnosis of crime reporter flies in forensic entomology: a review. Indian J Entomol 80(2):158–176

Gibbs A, Kuenzli M, Blomquist GJ (1995) Sex- and age-related changes in the biophysical properties of cuticular lipids of the housefly, Musca domestica. Arch Insect Biochem Physiol 29(1):87–97

GilArriortua M, Bordas MI, Ciné LM, Pinheiro F, de Pancorbo MM (2013) Cytochrome b as a useful tool for the identification of blow flies of forensic interest (Diptera, Calliphoridae). Forensic Sci Int 228(1–3):132–136

Gomes L, Von Zuben CJ (2006) Forensic entomology and main challenges in Brazil. Neotrop Entomol 35(1):1–11

Gordon JM, Šobotník J, Chouvenc T (2020) Colony-age-dependent variation in cuticular hydrocarbon profiles in subterranean termite colonies. Ecol Evol 10(18):10095–10104

Griffiths AM, Evans LM, Stevens JR (2009) Characterization and utilization of microsatellite loci in the New World screwworm fly, Cochliomyia hominivorax. Med Vet Entomol Suppl 1:8–13

Grzywacz A, Ogiela J, Tofilski A (2017) Identification of Muscidae (Diptera) of medico-legal importance by means of wing measurements. Parasitol Res 116(5):1495–1504

Guo YD, Cai JF, Xiong F, Wang HJ, Wen JF, Li JB, Chen YQ (2012) The utility of mitochondrial DNA fragments for genetic identification of forensically important sarcophagid flies (Diptera: Sarcophagidae) in China. Trop Biomed 29(1):51–60

Hall RD (1990) Medicocriminal entomology. In: Catts EP, Haskell NH (eds) Entomology and death: a procedural guide||. Joyce's Print Shop, Clemson, SC, pp 1–8

Hall RD (2001) Introduction: perceptions and status of forensic entomology. In: Byrd JH, Castner JL (eds) Forensic entomology: the utility of arthropods in legal investigations. CRC Press, Boca Raton, FL, pp 1–15

Hall MJR (2021) The relationship between research and casework in forensic entomology. Insects 12(2):174

Harvey ML (2006) A molecular study of the forensically important calliphoridae (diptera): implications and applications for the future of forensic entomology. University of Western Australia

Harvey ML, Gaudieri S, Villet MH, Dadour IR (2008) A global study of forensically significant calliphorids: implications for identification. Forensic Sci Int 177:66–76

Høye TT, Ärje J, Bjerge K, Hansen OLP, Iosifidis A, Leese F, Mann HMR, Meissner K, Melvad C, Raitoharju J (2021) Deep learning and computer vision will transform entomology. Proc Natl Acad Sci U S A 118(2):e2002545117

Kavitha R, Nazni WA, Tan TC, Lee HL, Isa MN, Azirun MS (2012) Molecular identification of blow flies recovered from human cadavers during crime scene investigations in Malaysia. Malays J Pathol 34:127–132

Kavitha R, Tan TC, Lee HL, Nazni WA, Sofian AM (2013a) Molecular identification of Malaysian Chrysomya megacephala (Fabricius) and Chrysomya rufifacies (Macquart) using life stage specific mitochondrial DNA. Trop Biomed 30(2):211–219

Kavitha R, Tan TC, Lee HL, Nazni WA, Sofian-Azirun M (2013b) DNA typing of Calliphorids collected from human corpses in Malaysia. Trop Biomed 30(1):119–124

Khang TF, Mohd Puaad NAD, Teh SH, Mohamed Z (2021) Random forests for predicting species identity of forensically important blow flies (Diptera: Calliphoridae) and flesh flies (Diptera: Sarcophagidae) using geometric morphometric data: Proof of concept. J Forensic Sci 66(3): 960–970

Kotzé Z, Aimar S, Amendt J, Anderson GS, Bourguignon L, Hall MJR, Tomberlin JK (2021) The forensic entomology case report—a global perspective. Insects 12(4):283

Lang MD, Allen GR, Horton BJ (2006) Blowfly succession from possum (Trichosurus vulpecula) carrion in a sheep-farming zone. Med Vet Entomol 20:445–452

Lee SF, Chen Z, McGrath A, Good RT, Batterham P (2011) Identification, analysis, and linkage mapping of expressed sequence tags from the Australian sheep blowfly. BMC Genomics 10:406

Lunas BM, de Paula MC, Michelutti KB, Lima-Junior SE, Antonialli-Junior WF, Cardoso CAL (2019) Hydrocarbon and fatty acid composition from Blowfly eggs represents a potential complementary taxonomic tool of forensic importance. J Forensic Sci 64(6):1720–1725

Luquin MFH, Santacruz EV, Morales RAL, Vázquez CN, Zúñiga MG (2017) Development of intelligent tools for recognizing cockroaches in the forensic entomology context. In: 2017 Intelligent Systems Conference (IntelliSys), pp 1117–1121

Magni PA, Harvey ML, Saravo L, Dadour IR (2012) Entomological evidence: lessons to be learnt from a cold case review. Forensic Sci Int 223(1–3):e31–e34

Matuszewski S (2021) Post-mortem interval estimation based on insect evidence: current challenges. Insects 12(4):314

McDonagh L, Thornton C, Wallman JF, Stevens JR (2009) Development of an antigen-based rapid diagnostic test for the identification of blowfly (Calliphoridae) species of forensic significance. Forensic Sci Int Genet 3(3):162–165

McKnight BE (1981) The washing away of wrongs: forensic medicine in thirteenth century China||, trans. S. Tzu., University of Michigan, Ann Arbor, pp. 196

Meiklejohn KA, Wallman JF, Dowton M (2011) DNA-based identification of forensically important Australian Sarcophagidae (Diptera). Int J Legal Med 125(1):27–32

Meng F, Ren L, Wang Z, Deng J, Guo Y, Chen C, Finkelbergs D, Cai J (2017) Identification of forensically important Blow flies (Diptera: Calliphoridae) in China based on COI. J Med Entomol 54(5):1193–1200

Mohsin, Kamshad (2021) Artificial intelligence in forensic science (August 24, 2021). Available at SSRN: https://ssrn.com/abstract=3910244 or https://doi.org/10.2139/ssrn.3910244)

Moore HE, Adam CD, Drijfhout FP (2013) Potential use of hydrocarbons for aging Lucilia sericata blowfly larvae to establish the postmortem interval. J Foren Sci 58(2):404–412

Moore HE, Adam CD, Drijfhout FP (2014) Identifying 1st instar larvae for three forensically important blowfly species using "fingerprint" cuticular hydrocarbon analysis. Foren Sci Int 240: 48–53

Moore HE, Butcher JB, Adam CD, Day CR, Drijfhout FP (2016) Age estimation of Calliphora (Diptera: Calliphoridae) larvae using cuticular hydrocarbon analysis and artificial neural networks. Foren Sci Int 268:81–91

Moore HE, Butcher JB, Day CR, Drijfhout FP (2017a) Adult fly age estimations using cuticular hydrocarbons and artificial neural networks in forensically important Calliphoridae species. Forensic Sci Int 280:233–244

Moore HE, Pechal JL, Benbow ME, Drijfhout FP (2017b) The potential use of cuticular hydrocarbons and multivariate analysis to age empty puparial cases of Calliphora vicina and Lucilia sericata. Sci Rep 7(1):1933

Moore HE, Hall MJR, Drijfhout FP, Cody RB, Whitmore D (2021) Cuticular hydrocarbons for identifying Sarcophagidae (Diptera). Sci Rep 11(1):7732

Moore H, Lutz L, Bernhardt V, Drijfhout FP, Cody RB, Amendt J (2022) Cuticular hydrocarbons for the identification and geographic assignment of empty puparia of forensically important flies. Int J Legal Med 136(6):1791–1800

Napoleão KS, Mello-Patiu CA, Oliveira-Costa J, Takiya DM, Silva R, Moura-Neto RS (2016) DNA-based identification of forensically important species of Sarcophagidae (Insecta: Diptera) from Rio de Janeiro, Brazil. Genet Mol Res 15(2):1–7

Negre B, Simpson P (2013) Diversity of transposable elements and repeats in a 600 kb region of the fly Calliphora vicina. Mob DNA 4:13

Nelson LA, Wallman JF, Dowton M (2008) Identification of forensically important Chrysomya (Diptera: Calliphoridae) species using the second ribosomal internal transcribed spacer (ITS2). Foren Sci Int 177:238–247

Owings CG, McKee-Zech HS, Schwing ST, Bugajski KN, Davis MC, Steadman DW (2022) Not by the book: observations of delayed oviposition and re-colonization of human remains by Blow flies. Insects 13(10):879

Park SH, Park CH, Zhang Y, Piao H, Chung U, Kim SY, Ko KS, Yi CH, Jo TH, Hwang JJ (2013) Using the developmental gene bicoid to identify species of forensically important blow flies (Diptera: calliphoridae). Biomed Res Int 2013:538051. https://doi.org/10.1155/2013/538051

Paula MC, Antonialli-Junior WF, Mendonça A, Michelutti KB, Eulalio AD, Cardoso CA, de Lima T, Von Zuben CJ (2017) Chemotaxonomic profile and intraspecific variation in the Blow Fly of Forensic Interest Chrysomya megacephala (Diptera: Calliphoridae). J Med Entomol 54(1):14–23

Pechal JL, Moore H, Drijfhout F, Benbow ME (2014) Hydrocarbon profiles throughout adult Calliphoridae aging: a promising tool for forensic entomology. Foren Sci Int 245:65–71

Picard CJ, Wells JD, Ullyot A, Rognes K (2017) Amplified fragment length polymorphism analysis supports the valid separate species status of *Lucilia caesar* and *L. illustris* (Diptera: Calliphoridae). Forensic Sci Res 3(1):60–64

Pickering CL, Hands JR, Fullwood LM, Smith JA, Baker MJ (2015) Rapid discrimination of maggots utilising ATR-FTIR spectroscopy. Foren Sci Int 249:189–196

Preativatanyou K, Sirisup N, Payungporn S, Poovorawan Y, Thavara U, Tawatsin A, Sungpradit S, Siriyasatien P (2010) Mitochondrial DNA-based identification of some forensically important blowflies in Thailand. Foren Sci Int 202:97–101

Roux O, Gers C, Legal L (2008) Ontogenetic study of three Calliphoridae of forensic importance through cuticular hydrocarbon analysis. Med Vet Entomol 22(4):309–317

Salem AM, Adham FK, Picard CJ (2015) Survey of the genetic diversity of forensically important chrysomya (Diptera: Calliphoridae) from Egypt. J Med Entomol 52(3):320–328

Sharma M, Singh D (2015) Historical approach of forensic entomology: a review. Int J Adv Res Sci Eng Technol 2:9

Sharma M, Singh D, Sharma AK (2015) Mitochondrial DNA based identification of forensically important Indian flesh flies (Diptera: Sarcophagidae). Forensic Sci Int 247:1–6

Sharma A, Drijfhout FP, Tomberlin JK, Bala M (2021) Cuticular hydrocarbons as a tool for determining the age of Chrysomya rufifacies (Diptera: Calliphoridae) larvae. J Forensic Sci 66(1):236–244

Sonet G, Jordaens K, Braet Y, Desmyter S (2012) Why is the molecular identification of the forensically important blowfly species Lucilia caesar and L. illustris (family Calliphoridae) so problematic? Foren Sci Int 223(1–3):153–159

Sontigun N, Sukontason KL, Amendt J, Zajac BK, Zehner R, Sukontason K, Chareonviriyaphap T, Wannasan A (2018) molecular analysis of forensically important blow flies in Thailand. Insects 9(4):159

Sperling FA, Anderson GS, Hickey DA (1994) A DNA-based approach to the identification of insect species used for postmortem interval estimation. J Forensic Sci 39(2):418–427

Stamper T, Wong ES, Timm A, DeBry RW (2017) Validating sonication as a DNA extraction method for use with carrion flies. Forensic Sci Int 275:171–177

Trabalon M, Campan M, Clement JL, Lange C, Miquel MT (1992) Cuticular hydrocarbons of Calliphora vomitoria (Diptera): relation to age and sex. Gen Comp Endocrinol 85(2):208–216

Wang Y, Wang Y, Wang M, Xu W, Zhang Y, Wang J (2021) Forensic entomology in China and its challenges. Insects 12(3):230

Wells JD, Sperling FA (2001) DNA-based identification of forensically important Chrysomyinae (Diptera: Calliphoridae). Forensic Sci Int 120(1–2):110–115

Wells JD, Stevens JR (2008) Application of DNA-based methods in forensic entomology. Annu Rev Entomol 53:103–120

Xu H, Ye GY, Xu Y, Hu C, Zhu GH (2014) Age-dependent changes in cuticular hydrocarbons of larvae in Aldrichina grahami (Aldrich) (Diptera: Calliphoridae). Forensic Sci Int 242:236–241

Yang L, Cai J, Wen J, Guo Y (2010) Sequencing of mitochondrial DNA cytochrome oxidase subunit I gene in sarcosaphagous flies from 14 provinces in China. Zhong Nan Da Xue Xue Bao Yi Xue Ban 35:819–825

Ye G, Li K, Zhu J, Zhu G, Hu C (2007) Cuticular hydrocarbon composition in pupal exuviae for taxonomic differentiation of six necrophagous flies. J Med Entomol 44(3):450–456

Yusseff-Vanegas SZ, Agnarsson I (2017) DNA-barcoding of forensically important blow flies (Diptera: Calliphoridae) in the Caribbean Region. PeerJ 5:e3516

Zaidi F, Wei SJ, Shi M, Chen XX (2011) Utility of multi-gene loci for forensic species diagnosis of blowflies. J Insect Sci 11:59

Zhang C, Fu X, Xie K, Yan W, Guo Y (2015) MtDNA Analysis for genetic identification of forensically important sarcophagid flies (Diptera: Sarcophagidae) in China. J Med Entomol 52(6):1225–1233

Zhang X, Shang Y, Ren L, Qu H, Zhu G, Guo Y (2022) A study of cuticular hydrocarbons of all life stages in Sarcophaga peregrina (Diptera: Sarcophagidae). J Med Entomol 59(1):108–119

Zhu GH, Ye GY, Hu C, Xu XH, Li K (2006) Development changes of cuticular hydrocarbons in Chrysomya rufifacies larvae: potential for determining larval age. Med Vet Entomol 20(4): 438–444

Zhu GH, Xu XH, Yu XJ, Zhang Y, Wang JF (2007) Puparial case hydrocarbons of Chrysomya megacephala as an indicator of the postmortem interval. Forensic Sci Int 169(1):1–5

Molecular Taxonomy and Forensic Entomology: Where We Stand Today

6

Abstract

Molecular taxonomy has become an essential tool today in forensic entomology in identifying museum and live specimens and identifying several life cycle stages and mature and immature forms. The importance of molecular taxonomy lay in identifying specimens that are otherwise difficult to be detected by morphology. We discuss in this chapter the different applications of molecular taxonomy in diverse species identification, in adult and immature forms and their application in forensic entomology.

Keywords

Molecular taxonomy · Forensic flies

Abbreviations

bp	Base pair
C albifrontalis	*Calliphora albifrontalis*
C dubia	*Calliphora dubia*
C graham	*Calliphora graham*
C macellaria	*Cochliomyia macellaria*
C pinguis (Walker),	*Chrysomya pinguis (Walker)*
C putoria	*Chrysomya putoria*
C rufifacies (Macquart)	*Chrysomya rufifacies (Macquart)*
C stygia	*Calliphora stygia*
C tibialis	*Conicera tibialis*
C varipes	*Chrysomya varipes*
C vicina	*Calliphora vicina*
C vomitoria	*Calliphora vomitoria*

S. Ghosh, D. Banerjee, *Biology of Forensically Important Invertebrates*,
https://doi.org/10.1007/978-981-99-5026-3_6

Ch megacephala (Fabricius)	*Chrysomya megacephala (Fabricius)*
COI	cytochrome oxidase subunit I
E peregrinum	*Euryomma peregrinum*
E philippinensis	*Exoristobia philippinensis*
F albitarsis	*Fannia albitarsis*
F collini	*Fannia collini*
F femoralis	*Fannia femoralis*
F fusconotata	*Fannia fusconotata*
F heydenii	*Fannia heydenii*
F nigra Malloch, 1010	*Fannia nigra* Malloch, 1910
F pallitibia Rondani, 1866	*Fannia pallitibia Rondani, 1866*
F sanihue	*Fannia sanihue*
FE	Forensic entomology
H aenescens	*Hydrotaea aenescens*
H ignava	*Hydrotaea ignava*
H illucens	*Hermetia illucens*
H ligurriens (Wiedemann)	*Hemipyrellia ligurriens (Wiedemann)*
H ligurriens	*Hemipyrellia ligurriens*
H rostrata	*Hydrotaea rostrata*
H similis	*Hydrotaea similis*
H violacea	*Hypopygiopsis violacea*
L ampullacea	*Lucilia ampullacea*
L bazini Séguy	*Lucilia bazini Séguy*
L caesar	*Lucilia caesar*
L cuprina (Wiedemann)	*Lucilia cuprina (Wiedemann)*
L eximia	*Lucilia eximia*
L hainanensis Fan	*Lucilia hainanensis Fan*
L illustris	*Lucilia illustris*
L porphyrina	*Lucilia porphyrina*
L sericata	*Lucilia sericata*
L silvarum	*Lucilia silvarum*
M domestica	*Musca domestica*
MIRS	Medium infra-red spectroscopy
NIRS	Near-infrared spectroscopy
O obesa	*Ornidia obesa*
P regina	*Phormia regina*
P terraenovae	*Protophormia terraenovae*
P vulgaris	*Parapiophila vulgaris*
PCR	Polymerase chain reaction
PMI_{min}	PMI minimum
r	Ribosomes
S albiceps	*Sarcophaga albiceps*
S caerulescens	*Sarcophaga caerulescens*
S dux	*Sarcophaga dux*
S impatiens	*Sarcophaga impatiens*

S martellata	*Sarcophaga martellata*
S nudiseta	*Synthesiomyia nudiseta*
S peregrina	*Sarcophaga peregrina*
S ruficornis	*Sarcophaga ruficornis*

6.1 Introduction

Forensic entomology is a science of application of insects collected in and around corpses and their identification to estimate the PMI in medicocriminal investigations and detection of crime (Harvey et al. 2016). Although mostly were identified using taxonomic keys, with the advances of molecular biological techniques, more and more emphasis is being laid on to application of molecular biology in identifying the insect specimens This finds more relevance in case of specimens which are damaged or found single or improperly preserved and identification by taxonomic keys is difficult. The study of mitochondrial gene cytochrome oxidase subunit I (COI) DNA, being the most popular markers with application in studies on molecular systematics, extracted from the insect specimens finds relevance in identification of the specimen. Spectroscopy is another such powerful tool that is enabling the study and taxonomic identification of forensically important flies.

A recent study on forensically important fly species has highlighted their distribution (Ren et al. 2018) (Table 6.1).

6.2 Molecular Taxonomy: Forensic Flies—A Global Study

Molecular genotyping methods of DNA-based specimen identification are another widely used method applied in the study of forensic entomology (Wells and Stevens 2008). Mitochondrial COI or cox1-based DNA barcoding approaches have been used to identify Fanniidae family from Asia, Europe, and North America for immature stages and poorly preserved specimens. (Grzywacz et al. 2017). DNA barcoding with the COI gene and a 1000-bp site of the 28S nuclear rRNA gene have enabled identification of forensic flies in specimens from Thailand, which were difficult to be identified by morphological tools (Zajac et al. 2016). Hermetia illucens (Linnaeus 1758), a European Stratiomyidae species, commonly termed as the black soldier fly, has been reported to feed on human corpses, and in a study reported from central Poland, centurion fly *Sargus bipunctatus* (Scopoli 1763) and its larvae were identified using morphological and DNA barcoding-based evidences. This also enabled distinguishing of *S. bipunctatus* and *H. Illucens* larva (Michalski et al. 2021).

Internal transcribed spacer 2 (ITS-2) region of nuclear ribosomal DNA (rDNA) or ITS2 has found importance as DNA barcodes due to important characteristics of conserved regions for universal primer designing, easy amplification, and variability

Table 6.1 Forensically important flesh flies and their distribution: an account[a]

No	Species	Location	Animal model	Habitat	References
1	*Boettcherisca highlandica* Kurahashi & Tan	Malaysia	Rabbits	Highland	Silahuddin et al. (2015)
2	*Blaesoxipha plinthopyga* (Wiedemann)	USA	Human	Mountain	Wells and Smith (2013)
3	*Liosarcophaga babiyari* (Lehrer)	Saudi Arabia	Rabbits	Mountain	Abouzied (2014)
4	*Oxysarcodexia intona* (Curran & Walley)	Brazil	Baited traps	Outdoor	de Sousa et al. (2015)
		Brazil	Pigs	Rainforest	Vasconcelos et al. (2013)
5	*Oxysarcodexia riograndensis* Lopes	Brazil	Human	Rural	Oliveira and Vasconcelos (2010)
		Brazil	Pigs	Rainforest	Vasconcelos et al. (2013)
6	*Oxysarcodexia thornax* (Walker)	Brazil	Baited traps	Outdoor	de Sousa et al. (2015)
		Brazil	Baited traps	Rural, urban, forest	de Souza and Von Zuben (2016)
7	*Peckia chrysostoma* (Wiedemann)	Brazil	Male cadaver	Indoor	Vasconcelos et al. (2014)
		Brazil	Baited traps	Outdoor	de Sousa et al. (2015)
8	*Peckia* (*Squamatodes*) *ingens* (Walker)	Brazil	Baited traps	Outdoor	de Sousa et al. (2015)
			Baited traps	Outdoor	Oliveira and Vasconcelos (2010)
			Pig	Rainforest	Vasconcelos et al. (2013)
9	*Peckia* (*Sarcodexia*) *lambens* (Wiedemann)	Brazil	Baited traps	Outdoor	de Sousa et al. (2015)
			Baited traps	Outdoor	Oliveira and Vasconcelos (2010)
			Baited traps	Rural, urban, forest	de Souza and Von Zuben (2016)
10	*Ravinia belforti* (Prado & Fonseca)	Brazil	Human corpse	Rural	Oliveira and Vasconcelos (2010)
11	*Ravinia pernix* (Harris)	Saudi Arabia	Rabbits		Mashaly (2016)

(continued)

Table 6.1 (continued)

No	Species	Location	Animal model	Habitat	References
				Agricultural/ desert/urban area	
12	*Sarcophaga aegyptiaca**Salem	Egypt	Rabbit	House	Abd El-bar and Sawaby (2011)
13	*Sarcophaga albiceps* Meigen	China	Pigs	Outdoor	Wang et al. (2008)
		China	Pigs	Outdoor	Chen (2000)
		India	Mutton	Wooden platform	Singh and Bharti (2008)
		Germany	Baited traps	Rural	Fremdt and Amendt (2014)
		Malaysia	Rabbits	Highland	Silahuddin et al. (2015)
		Poland	Pig	Forest and grassland	Szpila et al. (2015)
		India	Rabbits	Campus area	Bharti and Singh (2003)
14	*Sarcophaga africa**_(Wiedemann)	Switzerland	Human	Indoor	Cherix et al. (2012)
		Spain	Carrion-baited traps	Urban	Baz et al. (2014)
		Poland	Pig	Forest and grassland	Szpila et al. (2015)
		Kuwait	Rabbits	Outdoor	Al-Mesbah et al. (2012)
15	*Sarcophaga argyrostoma**_ (Robineau-Desvoidy)	Switzerland	Human	Indoor	Cherix et al. (2012)
		Poland	Pig	Forest and grassland	Szpila et al. (2015)
		Spain	Carrion-baited traps	Urban	Baz et al. (2014)
		Central Europe	Woman	Indoor	Pastula and Merritt (2013)
16	*Sarcophaga caerulescens**_ Zetterstedt	Southern Finland	Human	Indoor	Pohjoismäki et al. (2010)
		Switzerland			Cherix et al. (2012)
		Germany	Baited traps	Rural	Fremdt and Amendt (2014)
		Poland	Pigs	Grassland	Matuszewski et al. (2016)

(continued)

Table 6.1 (continued)

No	Species	Location	Animal model	Habitat	References
		Poland			Szpila et al. (2015)
17	*Sarcophaga carnaria* (Linnaeus)	Germany	Baited traps	Rural	Fremdt and Amendt (2014)
18	*Sarcophaga crassipalpis** Macquart	Spain	Carrion-baited traps	Indoor	Baz et al. (2014)
		Australia	Human	Indoor	Farrell et al. (2015)
		China	Man, pig and rabbit	Forest	Wang et al. (2017)
		Japan	Human	Unstated	Toukairin et al. (2017)
		Moravia	Woman	Indoor	Povolny and Verves (1997)
19	*Sarcophaga cultellata* Pandelle	Spain	Human corpse	Unstated	Velásquez et al. (2010)
20	*Sarcophaga dux* Thomson	Japan	Human	Unstated	Toukairin et al. (2017)
		Switzerland	Human	Outdoor	Cherix et al. (2012)
		Northern Thailand	Baited traps	Outdoor	Sukontason et al. (2010)
21	*Sarcophaga hirtipes* Wiedemann	India	Mutton	Wooden platform	Singh and Bharti (2008)
		India	Rabbits	Campus area	Bharti and Singh (2003)
		Saudi Arabia	Rabbits	Agricultural/desert/urban area	Mashaly (2016)
22	*Sarcophaga impatiens** Walker	Australia	Human	Indoor	Farrell et al. (2015)
23	*Sarcophaga melanura* Meigen	Poland	Pig	Forest and grassland	Szpila et al. (2015)
		Spain	Carrion-baited traps	Peri-urban	Baz et al. (2014)
24	*Sarcophaga peregrina** (Robineau-Desvoidy)	China	Pigs	Indoor	Chen (2000)
			Human	River	Liu et al. (2013)
			Man, pig and rabbit	Forest	Wang et al. (2017)
		Malaysia	Rabbits	Rural	

(continued)

Table 6.1 (continued)

No	Species	Location	Animal model	Habitat	References
					Silahuddin et al. (2015)
		Japan	Human	Unstated	Toukairin et al. (2017)
		Northern Thailand	Baited traps	Outdoor	Sukontason et al. (2010)
25	*Sarcophaga praedatrix* Walker	Australia	Human	Grassland	Farrell et al. (2015)
26	*Sarcophaga princeps* Wiedemann	Malaysia	Human, rabbits	Outdoor	Kumara et al. (2012)
					Tomberlin and Benbow (2015)
		India	Rabbits	Campus area	Bharti and Singh (2003)
27	*Sarcophaga ruficornis**_(Fabricius)	Australia	Human	Indoor	Farrell et al. (2015)
		Malaysia			Kumara et al. (2012)
		China	Pigs	Outdoor	Wang et al. (2008)
		Kuwait	Rabbits	Outdoor	Al-Mesbah et al. (2012)
		Northern Thailand	Baited traps	Outdoor	Sukontason et al. (2010)
28	*Sarcophaga similis**_Meade	Switzerland	Human	Indoor	Cherix et al. (2012)
		Poland	Pigs	Grassland	Oliveira and Vasconcelos (2010)
		Poland	Pigs	Forest and grassland	Szpila et al. (2015)
		Germany	Baited traps	Urban	Fremdt and Amendt (2014)
29	*Sarcophaga subvicina* Baranov	Germany	Baited traps	Rural	Fremdt and Amendt (2014)
30	*Sarcophaga taenionota* Wiedemann	China	Pigs	Outdoor	Wang et al. (2008)
		Malaysia	Rabbits	Rural/ highland	Silahuddin et al. (2015)
31	*Sarcophaga tibialis**_Macquart	Spain	Carrion-baited traps	Indoor	Baz et al. (2014)
32	*Sarcophaga variegata* (Scopoli)	Germany	Baited traps	Rural	Fremdt and Amendt (2014)
33	*Sarcophaga* spp.	Malaysia	Human	Indoor	Syamsa et al. (2010)

(continued)

Table 6.1 (continued)

No	Species	Location	Animal model	Habitat	References
				Indoor	Kumara et al. (2012)
				Indoor (high-rise buildings)	Syamsa et al. (2015)
				Indoor	Syamsa et al. (2017)
		Italy		Indoor	Bugelli et al. (2015)
34	*Tricharaea occidua* (Fabricius)	Brazil	Baited traps	Outdoor	de Sousa et al. (2015)
35	*Wohlfahrtia nuba* (Wiedemann)	Kuwait	Rabbits	Outdoor	Al-Mesbah et al. (2012)

[a] Adapted from Ren et al. 2018, Open Access article distributed under the terms of the Creative Commons Attribution License (http://creativecommons.org/licenses/by/4.0/)

in distinguishing closely related species. PCR-based studies using the ITS2 gene have identified 11 Calliphoridae and 5 Sarcophagidae fly species from Korea (Park et al. 2018). Mitochondrial DNA from whole blow fly and legs or from egg, larva, puparium, and empty puparium and its life cycle stages were used to study fly identification of insect and stage using DNA barcoding approaches in a study from Malaysia (Kavitha et al. 2013a, b). In forensic entomology studies, different baits have been used to attract and study the incumbent fly population. Pork bait has been used to attract adult and immature specimens revealing blow fly species of Lucilia and Calliphora and non-Calliphorid genera *Sarcophaga* and *Muscina* from Connecticut (Pinto et al. 2021). Using COI gene DNA barcoding approaches, PCR-based studies, and Sanger sequencing, multiplex real-time PCR system has been used to identify Calliphoridae and Sarcophagidae from South Korea (Jang et al. 2019).

DNA barcoding-based approaches have been used in species and specimen early developmental stages of eggs, first instar, and empty pupal cases enabled identification of Calliphoridae and Sarcophagidae families (Olekšáková et al. 2018). In the next section, we highlight few studies on different insect species through molecular taxonomy.

6.2.1 Calliphoridae

A study on Calliphoridae from Taiwan in the year 2000 has been identified using molecular data and COI-based approaches and aided in identification of *Hemipyrellia ligurriens (Wiedemann), Lucilia bazini Séguy, L. cuprina (Wiedemann), L. hainanensis Fan, L. prophyrina, Chrysomya megacephala (Fabricius)* or *Ch. megacephala (Fabricius), Chrysomya pinguis (Walker) or Ch. pinguis (Walker),* and *Chrysomya rufifacies (Macquart)* (Chen et al. 2004).

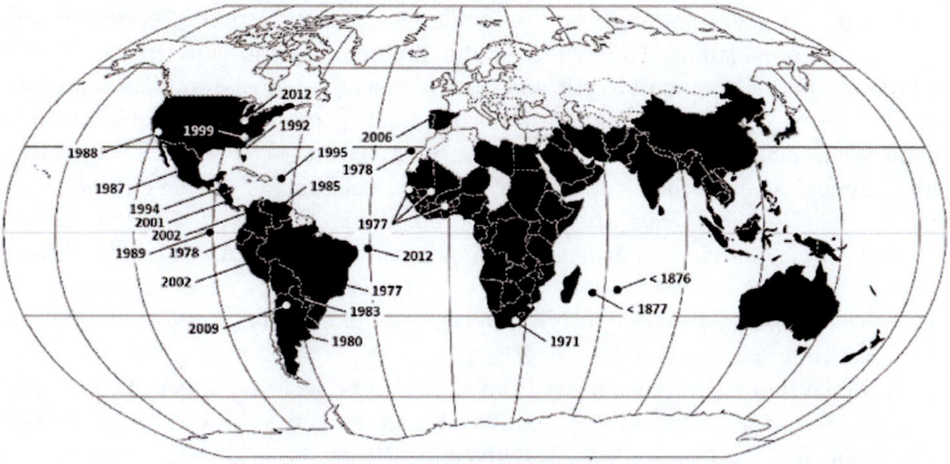

Fig. 6.1 Geographical distribution of *C megacephala*, reflecting the country and year of discovery outside native range (adapted from open-access article Badenhorst and Villet 2018)

COI has been used to identify species of Calliphoridae genera from Central Europe including *Calliphora* (C.) *vicina*, *Phormia (P.) regina*, and *Lucilia (L.) caesar, L. sericata*, and *L. ampullacea* (GilArriortua et al. 2014). Phylogenetic relationships within blow fly (Calliphoridae) species were studied using data from DNA sequences from the large subunit (lsu, 28S) ribosomal RNA (rRNA) gene, in flies studied from Britain and Europe (Stevens and Wall 2001).

Ch. megacephala, commonly called as oriental latrine fly, is predominant in distribution in Oriental and Australasian regions and finds importance as feed for livestock and in traditional Chinese medicine but suffers from challenges of identification of larvae and fragments of adults. COI barcode region analysis of *Protophormia (P.) terraenovae, C. megacephala*, and *Ch. pinguis* in flies studied from China revealed *C. megacephala* and *C. pinguis* as close relatives (Qiu et al. 2017). A total of 119 specimens of Calliphoridae species across different countries (22) were studied using 1167 bp of the COI gene. These included *L. cuprina, Ch. Albifrontalis, Ch. megacephala, Ch. saffranea*, and *Calliphora stygia* (Harvey et al. 2008). On the other hand, study conducted on *Ch. albiceps (Wiedemann), Ch. megacephala (F.)*, and *Ch. marginalis (Wiedemann),* under Calliphoridae from Egypt for COI gene, revealed interspecific divergence of 5.3% between *Ch. marginalis* and *Ch. megacephala*, 7% between *Ch. albiceps* and *Ch. megacephala*, and 8% between *Ch. albiceps* and *Ch. marginalis* (Salem et al. 2015). *Ch. megacephala* and *Ch. rufifacies* from Malaysia were identified by DNA sequencing of COI gene (Kavitha et al. 2013a, b). Genome size estimated for *Ch. rufifacies* (females) of 425.8 Mb and 1197.4 Mb *Haematobia irritans* (males) enabled determination of the species and sex of immature (Picard et al. 2012).

Very recently, R Badenhorst and MH. Villet have traced the global distribution of Ch. *megacephala*, and reports of molecular taxonomy-based identification have been growing (Fig. 6.1).

Using COI genes suffers from the limitation of identification of closely related species and populations. To circumvent this problem, a multi-gene approach using COI, cytochrome b subunit (CytB gene, being one of the most important mitochondrial genes used in identification of forensic insect species), NADH dehydrogenase subunit 5 (ND5), and nuclear internal transcribed spacers (ITS1 and ITS2) were used to study and analyze flies from China and Pakistan such as *Ch. megacephala F., Ch. pinguis Walker, L. sericata Meigen, L. porphyrina Walker, L. illustris Meigen, Hemipyrellia ligurriens Wiedemann, Aldrichina grahami Aldrich,* and *Musca domestica L.* under *Muscidae.* COI, CytB, and ITS2 were sequenced for *L. cuprina, Ch. megacephala, Ch. rufifacies,* and the flesh fly, *Sarcophaga albiceps Meigen*, from Pakistan (Zaidi et al. 2011).

In Malaysia maggot specimens from crime scene were identified using t-RNA leucine genes and cytochrome oxidase I and II genes in *Ch. megacephala, Ch. Rufifacies*, and *Ch. nigripes* (Kavitha et al. 2012).

COI barcode, CytB, 307 bp and the ITS2 (310–331 bp) has been used for the identification of blow flies from Southern European including *Calliphora species* including *(C. vicina, C. albiceps and C. vomitoria),* and *Lucilia (L. ampullaceal, L. Caesar, L. sericata* and *L. illustris)* which are predominant PMI indicators in Portugal (GilArriortua et al. 2015).

Ch. megacephala was identified using amplified fragment length polymorphism (AFLP) revealing two clusters in the population (Bao and Wells 2014). South African calliphorid species sequenced over an 1167 bp of the COI gene enabled species identification (Harvey et al. 2003a, b).

Quality of species identification has been improvised by molecular methods and samples collected from Germany including *Calliphora vicina, Calliphora vomitoria, L. caesar, L. sericata, L. illustris*, and *P terraenovae* which were identified to species level with a 229 bp sequence within the mitochondrial COI region (Reibe et al. 2009).

Ch. albiceps Wiedmann, Musca domestica L., M. albina Wiedemann, M. lucidula Loew, M. calleva Walker, M. sorbens Wiedmann, and *Physiphora alceae Preyssler* from Saudi Arabia were identified by DNA barcoding and sequencing of 710-bp Folmer region of COI gene (Mashaly et al. 2017). DNA barcoding using COI gene has been used to identify species from Germany including Calliphoridae: *Calliphora vicina, Calliphora vomitoria, Phormia regina, Protophormia terraenovae; Piophilidae: Parapiophila vulgaris*; *L. ampullacea, L caesar, L illustris, L sericata, L silvarum,* and Muscidae: *Hydrotaea* species including *H. dentipes, H. similis*, and *H. ignava* (Boehme et al. 2012).

COI and COII region-specific fragment amplification followed by digestion with different restriction enzymes (RE) like DraI recognising recognizes TTT^AAA sites or HinfI recognising recognizes G^ANTC sites revealed a 1.3 kb fragment, which has been used to identify *La sericata, C vicina* and *C vomitoria* (Schroeder et al. 2003).

PCR, followed by direct DNA sequencing, has been used to identify *Phormia regina, Phaenicia sericata*, and *L. illustris* (Sperling et al. 1994).

Blow flies, of subfamily Luciliinae, under Calliphoridae have been reported to be the first insects to colonize and produce eggs on decaying corpses in Korea. Using amplified and sequencing of the full length of the COI gene, the Luciliinae fly species were used to identify insect samples collected from Korea (Park et al. 2009).

DNA sequencing was used to identify species from southeastern Australia: *Chrysoma rufifacies, Ch. varipes, Calliphora albifrontalis, C. augur, C. dubia, C. hilli hilli, C. maritima, C. stygia, C. vicina*, and *Onesia tibialis*. All are known as carrion breeders, while *O. tibialis* is known as parasitoid in earthworms (Wallman and Donnellan 2001).

A 304-bp COI gene fragment, followed by RT-PCR and sequencing approaches, has been used to identify forensically important Diptera from China including Calliphoridae species *Ch. megacephala, Ch. albiceps, Ch. rufifacies, Ch. nigripes, Aldrichina grahami; Lucilla species including L. sericata, L caesar, L. cuprina, L. porphyrina,* and *L. bazini; Muscidae (M*) species including *M. autumnalis, M. domestica, Fannia canicularis, Stomoxys calcitrans*, and *Synthesiomyia nudiseta; Sarcophagidae (S) species* including *S. argyrostoma S. albiceps, S. dux, and Helicophagella melanura*; and *Calliphoridae* species including *Ch. megacephala, Ch. albiceps,* and *L. sericata* (Aly and Wen 2013).

COI studies have helped in identification of Calliphorid species from corpses in western Australia: *C. dubia, C augur*, *Ch. Rufifacies, L sericata*, and *Ch megacephala* (Harvey et al. 2003a, b).

Ch. pinguis (Walker), *L. porphyrina* (Walker), and *Ch. villeneuvi* (Patton) from Thailand were studied from human corpse based on sequencing of 1247 bp CO1 gene of the larvae and puparia (Monum et al. 2017). Sequencing of a 272-bp barcode fragment of COI could enable species-level identification of Chinese sarcophagid species (Guo et al. 2012a, b).

6.2.2 Sarcophagidae

Challenges in forensic entomology lay in identification of cryptic or poorly conserved specimens, and difficulty has been reported to be associated with identification of Sarcophagidae females. This has been circumvented by the application of sequencing of COX1 mitochondrial gene. This has led to identification of eight species in Sarcophagidae when *Sarcophaga argyrostoma* and *Sarcophaga tibialis* (Fuentes-López et al. 2021). Identification based on morphology of Sarcophagidae species suffers from limitations of availability of taxonomic expertise and limited forensic indicators. COI gene has been used to identify synanthropic flesh flies of *Peckia*, *Oxysarcodexia*, *Ravinia*, and *Tricharaea* genera from Colombia (Buenaventura et al. 2018).

COI-COII has been used to identify sequences of Iranian Sarcophagidae (Jafari et al. 2019). COI gene sequences have been used to identify Muscidae and Sarcophagidae in Korea by PCR amplification and automatic sequencing in *Sarcophaga harpax, Sarcophaga haemorrhoidalis, Hydrotaea occulta, Ophyra leucostoma, Muscina angustifrons, Muscina pascuorum*, and *Phaonia aureola* (Kim et al. 2014).

COI gene of European Sarcophaga was identified to species level by using the barcode region of 658 bp or using the entire COI region of 1535 bp (Jordaens et al. 2013a, b). Sarcophagidae were identified by using partial COI and COII genes for specimens of Egypt and China including *S. albiceps (Meigen), S. argyrostoma (Robineau-Desvoidy), S. dux (Thomson)*, and *Wohlfahrtia nuba (Wiedemann).* PCR amplification followed by direct sequencing of the amplification products and use of restriction enzymes HinfI and MfeI was used to study these flies. 700-bp COI fragment and 678 bp were used to identify Sarcophagidae specimens from China (Guo et al. 2014).

Sarcophagidae females and larvae were identified using DNA barcode 189-bp fragment of COII gene from specimens from China including *Boerttcherisca peregrina (Robineau-Desvoidy 1830), Parasarcophaga albiceps (Meigen 1826),* and *Parasarcophaga dux (Thompson 1869)* under *Sarcophagidae* (Guo et al. 2010).

A recent study has revealed DNA marker/tool-based identification of sarcophagid flies (Table 6.2).

6.2.3 Studies on Different Forensically Important Insects

COI sequences and the BOLD online database, or the Barcode of Life Data System (http://www.barcodinglife.org, Ratnasingham and Hebert 2007), used for acquisition, analysis, storage, and publication of DNA barcode records were used to identify Calliphoridae and Muscidae from baits of wildlife carcasses used as traps in a study from Portugal which has been identified by DNA barcoding-based approaches (Rolo et al. 2013). COI gene sequences of six Dermestes species collected from South Korea were identified by using PCR and sequencing-based approaches (Jang et al. 2019).

PCR followed by DNA sequencing of the amplification products and the analysis of the COII gene was used to identify blow fly species (Ying et al. 2007). *Synthesiomyia nudiseta* under Muscidae was reported from forensic cases in Spain.

Hymenoptera and forensically important ants find importance in colonization of a corpse but suffer from challenges in their morphological identification. Mitochondrial DNA was used to study specimens from rabbit carcass, with a 251-bp fragment of the 16S ribosomal gene RNA (16S rRNA) from Saudi Arabia (Alajmi et al. 2020).

Insect DNA barcoding technique has been improvised using PCR-based approaches as fast and effective method for insect species identification used in Mantodea, Phasmatodea, Neuroptera, Odonata, Blattodea, and Orthoptera orders and species under Lepidoptera, Coleoptera, Diptera, Hemiptera, and Hymenoptera (Thongjued et al. 2019).

Overcoming the limitations of available keys to European Fanniidae in identification of *Fannia* Robineau-Desvoidy, 1830, larvae (third instar) and females isolated from pig carrion, COI-based barcode studies proved effective in describing

Table 6.2 DNA-based identification of forensically important sarcophagid flies

No	DNA region	Amplified fragment length (bp)	Primer ID and sequences	Collection location	References
1	*COI*	783	Unstated	USA	Wells et al. (2001)
2	*COI*	278	C1-J-2495: 5′-CAGCTACTTTATGAGCTTTAGG-3′ C1-N-2800: 5′-CATTTCAAGCTGTGTAAGCATC-3′	Australia	
3	*COI*	304	5′-CTGCTACTTTATGAGCTTTAGG-3′ 5′-GATGCTTACACAACTTGAAATG-3′	Japan	Zehner et al. (2004)
4	*COI*	658	5′-GGTCWACWAATCATAAAGATATTGG-3′ 5′-RAAACTTCWGGRTGWCCAAARAATCA-3′	Australia	
5	*COI*	304	5′-CAGCTACTTTATGAGCTTTAGG-3′ 5′-CATTTCAAGCTGTGTAAGCATC-3′	China and Egypt	Aly and Wen (2013)
6	*COI*	127/658	TY-J-1460: TACAATTTATCGCCTAAACTTCAGCC C1-N-2191: CCCGGTAAAATTAAAATATAAACTTC C1-J-2183: CAACATTTATTTTGATTTTTTGG TL2-N-3014: TCCAATGCACTAATCTGCCATATTA 5′-AAAATTATAATAAARGCRTGRGC-3′ 5′-TCYACTAATCATAAAGATATTGGYAC-3′	West Europe	Jordaens et al. (2013a, b)
7	*COI*	272/1173	272- COI: 5′-CAGATCGAAATTTAAATACTTC-3′ 5′-GTATCAACATCTATTCCTAC-3′ 1173- COI: 5′-TACAATTTATCGCCTAAACTTCAGCC-3′ 5′-CAGCTACTTTATGAGCTTTAGG-3′	Egypt and China	Aly (2014)
8	*COI*	465	5′-CAGCTACTTTATGATCTTTAGG-3′ 5′-CATTTCAAGCTGTGTAAGCATC-3′	India	Sharma et al. (2015)
9	*COI*	400	Unstated	Brazil	Napoleão et al. (2016)
10	*COI* + *ND5*	296 + 386	COI: 5′-CAGCTACTTTATGATCTTTAGG-3′ COI: 5′-CATTTCAAGCTGTGTAAGCATC-3′	Germany	Zehner et al. (2004)

(continued)

Table 6.2 (continued)

No	DNA region	Amplified fragment length (bp)	Primer ID and sequences	Collection location	References
			ND5: 5′-CCAAAATATTCTGATCATCCTTG-3′ ND5: 5′-GGATTAACTGTTTGTTATACTTTTCG-3′	India	Bajpai and Tewari (2010)
11	*COI* + 16SrDNA	278 + 289	C1-J-2495: 5′-CAGCTACTTTATGAGCTTTAGG-3′ C1-N-2800: 5′-CATTTCAAGCTGTGTAAGCATC-3′ 5′-CGCTGTTATCCCTAAGGTAA-3′ 5′-CTGGTATGAAAGGTTTGACG-3′	China	Guo et al. (2011)
12	*COI* + period	700 + 678	COI: 5′-CTTTACCTGTACTTGCTGGAG-3′ COI: 5′-AACTTGTCGTTGTGATGCT-3′period: 5′-CGCTGTTATCCCTAAGGTAA-3′ period: 5′-CTGGTATGAAAGGTTTGACG-3′	China	Guo et al. (2014)
13	*COI* + 28SrDNA	Unstated	Unstated	Thailand (Chiang Mai)	Zajac et al. (2016)
14	*COII*	189	5′-ATTAGATGTTGATAATCG-3′ 5′-ACAAATTTC-TGAACATTG-3′	China	Guo et al. (2010)
15	*COII*	635	5′-AGAGCCTCTCCTTTAATAGAACA-3′ 5′-GAGACCATTACTTGCTTTCAGTCATC-3′	Egypt and China	Aly and Mahmoud (2016)
16	*COII* + 16S rDNA	637 + 555	C2-J-3138: 5′-AGAGCCTCTCCTTTAATAGAACA-3′ TK-N-3775: 5′-GAGACCATTACTTGCTTTCAGTCATC-3′ LR-J-12887: 5′-CCGGTCTGAACTCAGATCACGT-3′ LR-N-13398: 5′-CGCCTGTTTAACAAAAACAT-3′	China	Guo et al. (2012a, b)
17	*COI* + *COII*	2300	TY-J-1460: 5′-TACAATTTATCGCCTAAACTTCAGCC-3′ C1-N-2800: 5′-CATTTCAAGCTGTGTAAGCATC-3′ C1-J-2495: 5′-CAGCTACTTTATGAGCTTTAGG-3′ TK-N-3775: 5′-GAGACCATTACTTGCTTTCAGTCATCT-3′	Malaysia	Guo et al. (2014)
				China	Zhang et al. (2015a, b)

18	*COI* + *COII*	1300	5′-CAGCTACTTTATGAGCTTTAGG-3′ 5′-GAGACCATTACTTGCTTTCAGTCATCT-3′	Egypt and China	Aly et al. (2013)
19	12S and 16SrDNA + *ITS*	1172 + 1500	mtD-33F: 5′-ATGTTTTTGTTAAACAGGCG-3′ mtD-12SR: 5′-AAACTAGGATTAGATACCCTATTAT-3′ 18SF-1975F: 5′-TAACAAGGTTTCCGTAGGTG-3′ 28SR-52R: 5′-GTTAGTTTCTTTTCCTCCCCT-3′	Malaysia	Roziah et al. (2015)
20	*ITS2*	Unstated	ITS2_F: 5′-TGCTTGGACTACATATGGTTG A-3′ ITS2_R: 5′-GTAGTCCCATATGAGTTGAGGTT-3′	China	Song et al. (2008)
21	MtSNP markers	<150	Unstated	China	Zhang et al. (2015a, b)

Adapted from Ren et al. (2018), Open Access article distributed under the terms of the Creative Commons Attribution License (http://creativecommons.org/licenses/by/4.0/)

the *Fannia nigra Malloch, 1910, Fannia pallitibia (Rondani*, 1866, third instar larval stage), and *Fannia collini* (females, Grzywacz et al. 2021).

DNA-based methods have been used to identify six species under family Phoridae, *Megaselia scalaris, Me. giraudii, Me. abdita, Me. rufipes, Conicera tibialis, and Puliciphora borinquenensis* based on 658-bp region of the COI gene (Boehme et al. 2010).

Molecular identification has been immensely helpful in identification of the immature forms which is otherwise very difficult. Third instars *of Ch. megacephala* larvae were identified using PCR and sequencing of gustatory receptor 1 gene (CmegGr1), from samples collected from rabbit carcasses (Ghazali et al. 2022) (Table 6.3).

6.2.4 Studies on Parasites

Parasites find importance in forensic entomology, and myiasis is reported to cause huge damage to animal life, livestock, and economy. Recently, DNA- and PCR-based methods are being used to study molecular identification and diagnosis, taxonomy, and phylogenesis and enable designing of vaccination strategies against myiasis-causing larva of Sarcophagidae, Calliphoridae, and Oestridae species (Stevens and Wall 2002). *Sarcophaga bullata*, a myiasis-causing forensic fly, and its developmental stages and reproduction have been studied using RNA-Seq-based analyses revealing sex- and developmental-specific gene sets (Martison et al. 2019). Molecular taxonomy has widespread application in detecting livestock parasites. *Cochliomyia Townsend, 1915* (screwworms) is a broadly distributed, blow fly species in America. *Cochliomyia hominivorax (Coquerel)* and *Cochliomyia macellaria (Fabricius)* are livestock parasites, while *Cochliomyia aldrichi Del Ponte* and *Cochliomyia minima Shannon* have been studied by taxonomy, mitochondrial COI, and the nuclear EF-1α, ITS2, and 28S rRNA-based approaches. (Harvey 2005).

6.3 Different Other Techniques Used in the Study of Forensically Important Insects

6.3.1 Morphological

Recent advances in morphological identification by study of wings find importance in forensic entomology. Blow flies of the genera *L. Robineau-Desvoidy* were identified at species level using wing morphometric analysis of *Lucilia* from Thailand including *L. cuprina*, *L. porphyrina*, *L. sinensis* and *L. illustris* and from Switzerland including L. *caesar* (Limsopatham et al. 2022). Wing measurements have been found to have importance in identification of necrophagous Diptera from European *Calliphoridae,* including *C. vomitoria, Ch. albiceps, Phormia regina, P. terraenovae*, and *L. sericata* (Szpila et al. 2019).

Table 6.3 Forensic entomology studied through molecular-based identification approaches[a]

	Species names	Family: Calliphoridae
1	*Lucilia cuprina* (Wiedemann, 1830) (Australian sheep blow fly)	1. 304-bp COI gene fragment sequencing Aly et al. (2013) 2. 1167 bp COI gene Harvey et al. (2008)
2	*Lucilia ampullaceal* (Villeneuve, 1922)	1. CytB locus 2. Developmental gene bicoid sequences used to identify 12 blow fly species, of Calliphoridae (Luciliinae, Chrysomyinae, and Calliphorinae) subfamilies Park et al. (2013)
3	*Lucilia caesar* (Linnaeus, 1758)	304-bp COI gene fragment Aly et al. (2013)
4	*Lucilia sericata* (Meigen, 1826)	1. Bicoid gene sequences used to study 12 blowfly species, under Calliphoridae (Luciliinae, Chrysomyinae, and Calliphorinae) subfamilies Park et al. (2013) 2. COI-based identification of three Calliphorid species (Harvey et al. 2003a, b, 2008)
5	*Lucilia silvarum* (Meigen, 1826)	COI gene barcoding region used to identify 111 specimens under 13 species of Calliphoridae from Germany (Boehme et al. 2012)
6	*Lucilia eximia* (Wiedemann, 1819)	A-T rich region in insects (Lessinger et al. 2000)
7	*Lucilia illustris* (Meigen, 1826)	1. Bicoid gene sequences used to identify Calliphoridae (Luciliinae, Chrysomyinae, and Calliphorinae) subfamilies (Park et al. 2013) 2. COI gene barcoding region used to identify 111 specimens under 13 species of Calliphoridae from Germany (Boehme et al. 2012)
8	*Calliphora graham*	Bicoid gene sequences (Park et al. 2013)
9	*Calliphora vomitoria* (Linnaeus, 1758)	Cytochrome b locus (Gilarriortua et al. 2013)
10	*Calliphora vicina* (Robineau-Desvoidy, 1830)	Bicoid gene sequences (Park et al. 2013)
11	*Calliphora stygia* (Fabricius, 1781)	1167 bp COI gene (Harvey et al. 2008)
12	*Calliphora dubia*	COI gene sequencing (Harvey et al. 2003a, b)
13	*Calliphora albifrontalis*	1167 bp of COI gene (Harvey et al. 2008)
14	*Chrysomya megacephala* (Fabricius, 1794)	1. Bicoid gene sequences (Park et al. 2013) 2. 304-bp COI gene fragment sequencing (Aly and Wen 2013) 3. mtDNA control region (A-T-rich region) (Lessinger et al. 2000)
15	*Chrysomya putoria* (Wiedemann, 1830) (African latrine fly)	Karyotypes, constitutive heterochromatin, and genomic DNA values
16	*Chrysomya rufifacies*	1. 304-bp COI gene fragment (Aly and Wen 2013) 2. COI gene sequencing (Harvey et al. 2003a, b)
17	*Chrysomya varipes*	Karyotypes, constitutive heterochromatin, and genomic DNA values

(continued)

Table 6.3 (continued)

	Species names	Family: Calliphoridae
18	*Hypopygiopsis violacea*	COI gene barcoding Encyclopedia of Life (2014)
19	*Cochliomyia macellaria* (Fabricius, 1775)	Microsatellite loci characterization (Balme et al. 2012)
20	*Protophormia terraenovae* (Robineau-Desvoidy, 1830)	1. Karyotypes, constitutive heterochromatin, and genomic DNA
22	*Hemipyrellia ligurriens* (Wiedemann, 1830)	Bicoid gene sequences (Park et al. 2013)
23	*Phormia regina* (Meigen, 1826)	Bicoid gene sequences (Park et al. 2013)
Family: Muscidae		
24	*Hydrotaea ignava* (Harris, 1780)	COI barcode reported in Germany)Boehme et al. 2012)
25	*Hydrotaea similis* (Meade, 1887)	COI barcode (Boehme et al. 2012)
26	*Fannia fusconotata*	COI barcode (Encyclopedia of Life 2014)
27	*Fannia albitarsis*	COI barcode (Encyclopedia of Life 2014)
28	*Fannia femoralis*	COI barcode (Encyclopedia of Life 2014)
29	*Fannia heydenii* (Wiedemann, 1830)	COI barcode (Encyclopedia of Life 2014)
30	*Hydrotaea aenescens* (Wiedemann, 1830) (black dump flies)	GC-MS analysis of puparia cuticular lipids (hydrocarbons, waxes). Puparia lipid composition showed changes over time (Frere et al. 2014)
31	*Hydrotaea rostrata* (Robineau-Desvoidy, 1830)	COI or COX1 barcoding (Encyclopedia of Life 2014)
32	*Synthesiomyia nudiseta*	304-bp COI gene fragment (Aly and Wen 2013)
33	*Musca domestica* (Linnaeus, 1758) Muscidae	304-bp COI gene fragment (Aly and Wen 2013)
Family: Sarcophagidae		
34	*Sarcophaga albiceps* (Meigen, 1826)	304-bp COI gene fragment (Aly and Wen 2013)
35	*Sarcophaga dux* (Thomson, 1869)	1. Partial mitochondrial COI and COII genes used to identify important Sarcophagidae from Egypt and China (Aly et al. 2013) 2. 189 bp of COII gene (Guo et al. 2010)
36	*Sarcophaga martellata* (Senior-White, 1924)	*S. martellata* female determined by DNA sequencing (COI and COII) and PCR-RFLP (COI) and compared with *S ruficornis* (Tan et al. 2010)
37	*Sarcophaga ruficornis* (Fabricius, 1914)	
38	*Sarcophaga impatiens* (Walker, 1849)	COI barcode sequences used to study immature life stage forensically important Australian flesh fly (Meiklejohn et al. 2013)
39	*Sarcophaga caerulescens* (Zetterstedt, 1838)	COI gene barcoding region (Encyclopedia of life 2014)
40	*Sarcophaga peregrina* (Robineau-Desvoidy, 1830)	1. COI and 289-bp of 16S rDNA regions sequenced 2. 14,922 bp circular mitochondrial genome with

(continued)

Table 6.3 (continued)

	Species names	Family: Calliphoridae
		13 protein-coding genes, 22 transfer RNA genes, 2 ribosomal RNA genes and contains one noncoding A-Trich region
Other relevant families		
Family: Fanniidae		
41	*Fannia sanihue* (Domínguez and Aballay, 2008)	COI gene "barcoding region" (Encyclopedia of life 2014)
42	*Euryomma peregrinum* (Meigen, 1826)	COI gene barcoding region (Encyclopedia of life 2014)
Family: Milichiidae		
43	*Desmometopa* sp.	COI gene barcoding region (Encyclopedia of life 2014)
Family: Phoridae		
44	*Conicera tibialis* (Scmitz, 1925) (coffin flies)	658-bp-long region of COI gene (Boehme et al. 2010)
Family: Piophilidae		
45	*Parapiophila vulgaris* (Fallen, 1820)	COI barcode used to identify forensically relevant Diptera in Germany (Boehme et al. 2012)
Family: Encyrtidae		
46	*Exoristobia philippinensis* (Ashmead, 1904)	COI gene barcoding region (Encyclopedia of life 2014)
Family: Stratiomyidae		
47	*Hermetia illucens* (Linnaeus, 1758) (Black soldier fly)	Two serine proteases (SPs) of the larvae revealed that Hi-SP1 may be a larval-specific chymotrypsin-like protease involved with food digestion, while Hi-SP2 may be a trypsin-like protease (Kim et al. 2011)
Family: Syrphidae		
48	*Ornidia obesa* (Fabricius, 1775)	COI gene barcoding region (Encyclopedia of life 2014)

[a] *PMI* Postmortem interval, *COI* cytochrome oxidase subunit I, *COII* cytochrome oxidase subunit II

Members of Muscidae that play a major role in PMI_{min} determination in forensic investigations suffer from disadvantages of limited taxonomic key. Molecular methods of wing geometric morphometry have helped identification in *M. domestica Linnaeus, M. pattoni Austen, M. ventrosa Wiedemann, Hydrotaea chalcogaster (Wiedemann), Hydrotaea spinigera Stein*, and *Dichaetomyia quadrata (Wiedemann)* (Limsopatham et al. 2021). Blow fly species of Thailand has been identified using wing shape variation within species and sexes in Chrysomya species (Sontigun et al. 2017).

External hyperspectral imaging (HSI) of dorsal and ventral sides of *Calliphora dubia* and *Ch. rufifacies* puparia has been reported to enable their identification and revealed to be a nondestructive method for species identification (Voss et al. 2017).

6.3.2 Infrared Spectroscopy (IR) Techniques

IR spectroscopy is another such recently developed technique that is being applied to forensic entomology aiding in the study of forensic entomology together with multivariate statistical analysis of data which has been applied in the field of study of entomology, taxonomy, and toxicology and is a developing field for a span of over two decades now. It has the advantage of being a fast, reliable, and nondestructive tool. Very recently, near-infrared spectroscopy (NIRS) and medium infrared spectroscopy (MIRS) have been applied to forensic entomology (FE) and taxonomy, in prediction of larval age, detection of toxic substance in insects taken from environments or dead bodies, internal or external infestations by live or dead insects, and detection in stored products (Jales et al. 2021).

6.3.3 Gut Contents

Recently, insect gut contents have been used to study the forensic entomology (Wells and Stevens 2008).

6.4 Studies on Indian Species of Forensic Flies

Calliphorid flies, distributed widely across zoo-geographical zones, find importance in agricultural sciences, medical, sanitary, veterinary, and forensic sciences.

The adult flies are known to transmit disease-causing pathogens including bacteria, fungi, viruses, helminths, and protozoans causing infectious diseases like diarrhea, amoebic and bacillary dysentery, plague, tuberculosis poliomyelitis, and cholera (Greenberg 1971), due to their breeding habits on dead bodies, garbage, excreta, and transmitting these pathogens to healthy individuals causing menace to public health.

The larvae are parasitic, predaceous, saprophagous or scavengers, and sanguinivorous. Parasitic larval are known to cause myiasis in humans and other animals (Dasgupta 1962; James 1947; Zumpt 1965).

Insect-mediated transmission of myasis has been reported to cause severe outbreak of cutaneous myiasis in sheep, reported to cause mass death of sheep (Kettle 1995).

The parasitic or predator larvae has been reported to affect other organisms including invertebrates like earthworms (anellids), snails (mollusks), and arthropods (locusts) and vertebrates like Amphibians such as toads, frogs, and Aves like nestling birds (Draber-M6nko 1981; Rognes 1992), causing loss of fish and meat products (Sinha and Nandi 2004; Esser 1991) and adversely affecting the food industry.

The family members of forensic flies including blow flies, bluebottles or greenbottles due to their blue, green, yellow, brown, gray, or black colors, reveal an oviparous nature with few viviparous ones.

Senior-White (1922, 1923, 1924a,d, 1925, 1926), Senior-White et al. (1940) had studied and reported in-depth about the Calliphoridae species from Indian subcontinent. Morphologically, they reveal several unique features in adults with well-developed thoracic squamae, notopleural (02) with a hypopleural bristles as distivt rows and the larvae with posterior spiracles in slight depression in posterior plate.

Very few works have been reported from different parts of India by different authors, and a complete record is still not available. Till 2004, around 112 species under 30 genera were reported which included from 35 political subdivsions of Indian subregion (Nandi 2004).

Although *Ch. saffranea (Bigot 1877)* is currently endemic and recorded only in Australia and New Guinea, in 2016, it was reported from Aurangabad City, Maharashtra State, in India and collected as full-grown third instar larvae collected from decaying cats. This larva when reared under laboratory conditions led to emergence of adult flies from pupae which were collected and identified by their morphological features (Abd-AlGalil et al. 2016).

Several studies from India have reported that after death, wounds and natural openings of host body are breeding place for flies. Here they stay and thrive on body fluids and consume the body very quickly and start and passed a life cycle. The blow fly life stages and the developmental time of an adult blow fly can help in PMI estimation (Kulshrestha and Chandra 1987). But due to the lack/incompleteness of suitable morphological keys for immature stages of forensic flies, COI reference data were developed from nine species under three subfamilies, Calliphorinae, Luciliinae, and Chrysomyinae, of blow flies from India (Bharti and Singh 2017).

The blow fly *Ch. megacephala* (Fabricius 1794) from India has been studied for mouthpart trichoid sensilla (Tr), basiconic sensilla (Ba), and microtrichia (Mr) on labellar lobes, haustellum, and maxillary palps of mouthparts by scanning electron microscopy (SEM)-based studies (Bhattacherjee et al. 2022).

Cuticular hydrocarbons (CHC) from Calliphoridae, analyzed by Gas chromatography (GC), enabled age estimation, in first, second, feeding and post-feeding third instar of *Ch. rufifacies* commonly called as the hairy maggot blow fly revealing 23 compounds with predominant alkanes of 9–33 C in length, alkenes, and methyl-branched alkanes (17%). All the HCs except pentadecane (C-15), hexadecane (C-16), and nonacosane (C-29) revealed significant expression differences throughout development of larvae. For first instars, nonane was the most abundant, while tricosane and pentacosane, respectively, were the notable HCs in second instars. Studies indicated a low to high molecular weight (mw) HCs with the larval age (Sharma et al. 2021a).

Volatile organic compounds (VOCs) from *Ch. rufifacies (Macquart)* were used to estimate the age by GC-MS analysis. Thirty-seven compounds revealed both quantitative and qualitative shift with the larvae and pupae age. Compounds, like butanoic acid, 2-methyl butanoic acid, 2-ethyl-1-hexanol, phenol, octadecanoic acid, 2-methyl propanamide, 2-methyl butanoic acid, and hexadecanoic acid, serve as indicator compounds of specific stages within *Ch. rufifacies* development (Sharma et al. 2021b). A 465 bp fragment of COI gene-based study enabled identification of several Sarcophagidae species from India (Sharma et al. 2015).

Insects isolated from a carcass of submerged pig (*Sus scrofa*) during various seasons in two studies revealed presence of insects of Aeshnidae, Chironomidae, Coenagrionidae, and Lestidae families in early floating stage (second stage). In the third stage of floating decay of decomposition Aeshnidae. Burrowing water beetles, hemipteran species, and coleopteran species were reported from later stages of decomposition (Dalal et al. 2020).

The blow fly *Hemipyrellia ligurriens* (Wiedemann 1830) (Diptera: Calliphoridae) was studied by morphology of sensilla of antenna of adult male and female of *H. ligurriens*. Antenna of *H. ligurriens* revealed three segments including scape, pedicel, and flagellum with dorsolaterally placed arista, microtrichia, and trichoid sensilla. Chaetic sensilla I (ChI) was reported on the scape, and chaetic sensilla I and II (ChI and ChII) and styloconic sensilla were observed on the pedicel. The flagellum revealed three types of sensilla including trichoid, basiconic, and coeloconic sensilla together with basiconic sensilla with multiporous surfaces revealing characteristic olfactory function (Sharma et al. 2015).

S. albiceps Meigen and *S. hirtipes Wiedemann* were reported for their larval deposition during nighttime in September 2005 (Singh and Bharti 2008).

Ultrastructure of larvae and pupae of *Parasarcophaga ruficornis*, a common flesh fly species in India, was studied using light microscopy (LM) and scanning electron microscopy (SEM) which revealed identifying features of cephalopharyngeal apparatus, cephalic segment, spines, pupal respiratory horns, and anterior and posterior spiracles (Singh et al. 2012).

Ophyra capensis (Wiedemann 1818) was studied by SEM revealing antenna ocellar region and compound eye of adult male and female *Ophyra capensis* revealing larger ocellar region covered with microtrichia. Ultrastructural analysis of the compound eye indicated larger anterofrontally located ommatidia as compared to the rest of the facets, with sexual dimorphism of size of the ommatidia. The antennal ultrastructure revealed sensilla (five types) and chaetic sensilla (two types) on the scape and pedicel and trichoid sensilla and basiconic sensilla (two types) on the flagellum along with numerous microtrichia. Both types of basiconic sensilla displayed a multiporous surface highlighted their olfactory function (Hore et al. 2018).

Insect faunal succession studied from carcass of rabbit carcasses in a study from Punjab state in India, from March 1997 to December 1999, revealed insect species under Diptera, Coleoptera, and Hymenoptera with five species of Calliphoridae, four of Sarcophagidae, ten of Muscidae, and one each from Anthomyiidae and Otitidae. Coleopteran families included members of Cleridae, Staphylinidae, Dermestidae, Tenebrionidae, Histeridae and Silphidae, Formicidae (Hymenoptera), and Lepidoptera (Bharti and Singh 2003). *Ch. putoria* (Wiedemann 1830) and *C. thanomthini* Kurahashi and Tumrasvin (1977) were recorded from altitude of 2000 m in the Himalaya (Bharti 2019), revealing their presence in higher altitudes.

Recently, a checklist of necrophagous flies from India included 349 species belonging to families (16) and genera (56) which were reported recently in 2022 (Kar et al. 2022).

6.5 Discussion

In this chapter we have tried to draw a pen picture of the global occurrence of forensically important flies and how molecular taxonomy is aiding in their identification and characterization of adults, immature ones, and different stages of life cycle and thereby contributing to estimation of PMI (Hore et al. 2017; Jang et al. 2020; Lo Pinto et al. 2017; Martinson et al. 2019; Meiklejohn et al. 2011; Otranto and Stevens 2002; Singh and Bala 2009; Singh and Bharti 2001; Yusseff-Vanegas and Agnarsson 2016). We also provide an account of the studies done on Indian forensic flies through morphological, molecular, and biochemical analysis. This finds importance as forensic flies hold economic importance in disease-causing agents, medical and veterinary, sanitary, forensic and agricultural sciences, and study of their biodiversity across the globe. The distinction between closely related forms and immature on the species remains the key challenge to the forensic entomologists. More research is being conducted in search of better molecular markers to distinguish between closely related species and precise identification of immature stages so that diagnosis of the fly, parasites, and larval stages and different stages of development and estimation of PMI may be most accurate. In the next chapter, we discuss the importance of forensic flies and their role in the veterinary world.

References

Abd El-bar MM, Sawaby RF (2011) A preliminary investigation of insect colonization and succession on remains of rabbits treated with an organophosphate insecticide in El-Qalyubiya governorate of Egypt. Forensic Sci Int 208:26–30

Abd-AlGalil FMA, Zambare SP, Mashaly AMA (2016) First record of Chrysomya saffranea (Diptera: Calliphoridae) of forensic importance in India. Trop Biomed 33(1):102–108. PMID: 33579147

Abouzied EM (2014) Insect colonization and succession on rabbit carcasses in southwestern mountains of the Kingdom of Saudi Arabia. J Med Entomol 51:1168–1174

Alajmi R, Haddadi R, Abdel-Gaber R, Alkuriji M (2020) Molecular phylogeny of Monomorium pharaonis (hymenoptera: Formicidae) based on rRNA sequences of mitochondrial gene. J Genet 99:28

Al-Mesbah H, Moffatt C, El-Azazy OM et al (2012) The decomposition of rabbit carcasses and associated necrophagous Diptera in Kuwait. Forensic Sci Int 217:27–31

Aly SM (2014) Reliability of long vs short coi markers in dentification of forensically important flies. Croat Med J 55:19–26

Aly SM, Mahmoud SM (2016) COII "long fragment" reliability in characterisation and classification of forensically important flies. Arch Med Sadowej Kryminol 66:95–105

Aly SM, Wen J (2013) Applicability of partial characterization of cytochrome oxidase I in identification of forensically important flies (Diptera) from China and Egypt. Parasitol Res 112(7):2667–2674

Aly SM, Wen J, Wang X (2013) Identification of forensically important Sarcophagidae (Diptera) based on partial mitochondrial cytochrome oxidase I and II genes. Am J Forensic Med Pathol 34(2):159–163

Badenhorst R, Villet MH (2018) The uses of *Chrysomya Megacephala* (Fabricius, 1794) (Diptera: Calliphoridae) in forensic entomology. Forensic Sci Res 3(1):2

Bajpai N, Tewari RR (2010) Mitochondrial DNA sequence based phylogenetic relationship among flesh flies of the genus Sarcophaga (Sarcophagidae: Diptera). J Genet 89:51–54

Balme GR, Denning SS, Cammack JA, Watson DW (2012) Blow flies (Diptera: Calliphoridae) survive burial: Evidence of ascending vertical dispersal. Forensic Sci Int 216(1–3):e1–e4. https://doi.org/10.1016/j.forsciint.2011.07.017. ISSN 0379-0738

Bao F, Wells JD (2014) Population genetic structure of an invasive forensically important insect. Electrophoresis 35(21–22):3193–3200

Baz A, Botías C, Martínvega D et al (2014) Preliminary data on carrion insects in urban (indoor and outdoor) and periurban environments in Central Spain. Forensic Sci Int 248:41–47

Bharti M (2019) New records of Chrysomya putoria and C. thanomthini (Diptera: Calliphoridae) from India, with a revised key to the known Indian species. J Threatened Taxa 11(1): 13188–13190

Bharti M, Singh D (2003) Insect faunal succession on decaying rabbit carcasses in Punjab. India J Forensic Sci 48(5):1133–1143

Bharti M, Singh B (2017) DNA-based identification of forensically important blow flies (Diptera: Calliphoridae) from India. J Med Entomol 54(5):1151–1156

Bhattacherjee R, De S, Sharma G, Ghosh S, Mishra S, Suman DS, Banerjee D (2022) Prevalence of mouthpart sensilla and protease producing symbiotic gut bacteria in the forensic fly Chrysomya megacephala (Fabricius, 1794): insight from foraging to digestion. Acta Trop 229:106380

Boehme P, Amendt J, Disney RH, Zehner R (2010) Molecular identification of carrion-breeding scuttle flies (Diptera: Phoridae) using COI barcodes. Int J Legal Med 124(6):577–581

Boehme P, Amendt J, Zehner R (2012) The use of COI barcodes for molecular identification of forensically important fly species in Germany. Parasitol Res 110(6):2325–2332

Buenaventura E, Valverde-Castro C, Wolff M, Triana-Chavez O, Gómez-Palacio A (2018) DNA barcoding for identifying synanthropic flesh flies (Diptera, Sarcophagidae) of Colombia. Acta Trop 182:291–297

Bugelli V, Forni D, Bassi LA et al (2015) Forensic entomology and the estimation of the minimum time since death in indoor cases. J Forensic Sci 60:525–531

Chen LS (2000) Experimental study on postmortem interval with the invasion of sarcosaphagous insects on cadavers in different environments. Chin J Forensic Med 15:157–160

Chen WY, Hung TH, Shiao SF (2004) Molecular identification of forensically important blow fly species (Diptera: Calliphoridae) in Taiwan. J Med Entomol 41(1):47–57

Cherix D, Wyss C, Pape T (2012) Occurrences of flesh flies (Diptera: Sarcophagidae) on human cadavers in Switzerland, and their importance as forensic indicators. Forensic Sci Int 220:1–3

Dalal J, Sharma S, Bhardwaj T, Dhattarwal SK, Verma K (2020) Seasonal study of the decomposition pattern and insects on a submerged pig cadaver. J Forensic Legal Med 74:102023

Dasgupta B (1962) On the myiasis of the Indian toad Bufo melanostictus. Parasitology 52:53–66

de Sousa JR, Carvalho-Filho Fda S, Esposito MC (2015) Distribution and abundance of necrophagous flies (Diptera: Calliphoridae and Sarcophagidae) in Maranhão northeastern Brazil. J Insect Sci 15:15

de Souza CR, Von Zuben CJ (2016) Synanthropy of Sarcophagidae (Diptera) in southeastern Brazil. Neotrop Entomol 45:637–641

Esser JR (1991) Biology of Chrysomyamegacephala (Diptera:Calliphoridae) and reduction of losses caused to the salted dried fish industry in south-east Asia. Bull Entomol Res 81:33–42

Farrell JF, Whittington AE, Zalucki MP (2015) A review of necrophagous insects colonising human and animal cadavers in south-East Queensland, Australia. Forensic Sci Int 257:149–154

Fremdt H, Amendt J (2014) Species composition of forensically important blow flies (Diptera: Calliphoridae) and flesh flies (Diptera: Sarcophagidae) through space and time. Forensic Sci Int 236:1–9

Fuentes-López A, Pedreño Sala A, Romera E, Galián J (2021) DNA-based and taxonomic identification of forensically important Sarcophagidae (Diptera) in southeastern Spain. Sci Justice 61(2):150–159

Ghazali SNA, Emelia O, Hidayatulfathi O, Syamsa RA (2022) Potential application of gustatory receptor 1 (CmegGr1) gene as a molecular marker for identification of Chrysomya megacephala (Diptera: Calliphoridae). Trop Biomed 39(2):226–230

GilArriortua M, Salona Bordas MI, Cainé LM, Pinheiro F, de Pancorbo MM (2013) Cytochrome b as a useful tool for the identification of blowflies of forensic interest (Diptera, Calliphoridae). Forensic Sci Int 228(1–3):132–136. https://doi.org/10.1016/j.forsciint.2013.02.037

GilArriortua M, Saloña Bordas MI, Köhnemann S, Pfeiffer H, de Pancorbo MM (2014) Molecular differentiation of central European blowfly species (Diptera, Calliphoridae) using mitochondrial and nuclear genetic markers. Forensic Sci Int 242:274–282

GilArriortua M, Saloña-Bordas MI, Cainé LM, Pinheiro F, de Pancorbo MM (2015) Mitochondrial and nuclear DNA approaches for reliable identification of Lucilia (Diptera, Calliphoridae) species of forensic interest from southern Europe. Forensic Sci Int 257:393–397

Greenberg B (1971) Flies and disease. Ecology, classification, and biotic associations, vol 1. Prince ton Univ Press, Princeton, New Jersey. 856 pp

Grzywacz A, Wyborska D, Piwczyński M (2017) DNA barcoding allows identification of European Fanniidae (Diptera) of forensic interest. Forensic Sci Int 278:106–114

Grzywacz A, Jarmusz M, Walczak K, Skowronek R, Johnston NP, Szpila K (2021) DNA barcoding identifies unknown females and larvae of fannia R.-D. (diptera: fanniidae) from carrion succession experiment and case report. Insects 12(5):381

Guo YD, Cai JF, Li X, Xiong F, Su RN, Chen FL, Liu QL, Wang XH, Chang YF, Zhong M, Wang X, Wen JF (2010) Identification of the forensically important sarcophagid flies Boerttcherisca peregrina, Parasarcophaga albiceps and Parasarcophaga dux (Diptera: Sarcophagidae) based on COII gene in China. Trop Biomed 27(3):451–460

Guo Y, Cai J, Chang Y et al (2011) Identification of forensically important sarcophagid flies (Diptera: Sarcophagidae) in China, based on COI and 16S rDNA gene sequences. J Forensic Sci 56:1534–1540

Guo YD, Cai JF, Meng FM, Chang YF, Gu Y, Lan LM, Liang L, Wen JF (2012a) Identification of forensically important flesh flies based on a shorter fragment of the cytochrome oxidase subunit I gene in China. Med Vet Entomol 26(3):307–313

Guo YD, Cai JF, Xiong F et al (2012b) The utility of mitochondrial DNA fragments for genetic identification of forensically important sarcophagid flies (Diptera: Sarcophagidae) in China. Trop Biomed 29:51–60

Guo Y, Zha L, Yan W, Li P, Cai J, Wu L (2014) Identification of forensically important sarcophagid flies (Diptera: Sarcophagidae) in China based on COI and period gene. Int J Legal Med 128(1): 221–228

Harvey ML (2005) An alternative for the extraction and storage of DNA from insects in forensic entomology. J Forensic Sci 50(3):627–629

Harvey ML, Dadour IR, Gaudieri S (2003a) Mitochondrial DNA cytochrome oxidase I gene: potential for distinction between immature stages of some forensically important fly species (Diptera) in Western Australia. Forensic Sci Int 131(2–3):134–139

Harvey ML, Mansell MW, Villet MH, Dadour IR (2003b) Molecular identification of some forensically important blowflies of southern Africa and Australia. Med Vet Entomol 17(4): 363–369

Harvey ML, Gaudieri S, Villet MH, Dadour IR (2008) A global study of forensically significant calliphorids: implications for identification. Forensic Sci Int 177(1):66–76

Harvey M, Gasz N, Voss S (2016) Entomology-based methods for estimation of postmortem interval. Res Rep Forensic Med Sci 6:1–9. https://doi.org/10.2147/RRFMS.S68867

Hore G, Maity A, Naskar A, Ansar W, Ghosh S, Saha GK, Banerjee D (2017) Scanning electron microscopic studies on antenna of Hemipyrellia ligurriens (Wiedemann, 1830) (Diptera: Calliphoridae)-a blow fly species of forensic importance. Acta Trop 172:20–28

Hore G, Saha GK, Banerjee D (2018) Sensory organs of forensically important fly Ophyra capensis (Wiedemann, 1818) (Diptera: Muscidae): a scanning electron microscopic study. Acta Trop 185:400–411

Jafari S, Oshaghi MA, Akbarzadeh K, Abai MR, Koosha M, Mohtarami F (2019) Identification of forensically important flesh flies using the cytochrome C oxidase subunits I and II genes. J Med Entomol 56(5):1253–1259

Jales JT, Barbosa TM, de Medeiros JR, de Lima LAS, de Lima KMG, Gama RA (2021) Infrared spectroscopy and forensic entomology: can this union work? A literature review. J Forensic Sci 66(6):2080–2091

James MT (1947) The flies that cause myiasis in man. 631 = 1–175, U. S. Dept Agric Misc Publication

Jang H, Shin SE, Ko KS, Park SH (2019) SNP typing using multiplex real-time PCR assay for species identification of forensically important blowflies and Fleshflies collected in South Korea (Diptera: Calliphoridae and Sarcophagidae). Biomed Res Int 2019:6762517

Jang H, Shin SE, Youm KJ, Karagozlu MZ, Kim CB, Ko KS, Park SH (2020) Molecular identification of Necrophagous Dermestes species in South Korea using cytochrome c oxidase subunit I nucleotide sequences (genus Dermestes). J Forensic Sci 65(1):283–287

Jordaens K, Sonet G, Richet R, Dupont E, Braet Y, Desmyter S (2013a) Identification of forensically important Sarcophaga species (Diptera: Sarcophagidae) using the mitochondrial COI gene. Int J Legal Med 127(2):491–504

Jordaens K, Sonet G, Richet R et al (2013b) Identification of forensically important Sarcophaga species (Diptera: Sarcophagidae) using the mitochondrial COI gene. Int J Legal Med 127:491–504

Kar O, Naskar A, Pramanik D, Mukherjee K, Sarkar S, Mukherjee A, Ghosh D, Sengupta J, Banerjee D (2022) An inventory of necrophagous flies from India. Uttar Pradesh Journal of Zoology 43(23):79–108. https://doi.org/10.56557/upjoz/2022/v43i233262

Kavitha R, Nazni WA, Tan TC, Lee HL, Isa MN, Azirun MS (2012) Molecular identification of blow flies recovered from human cadavers during crime scene investigations in Malaysia. Malays J Pathol 34(2):127–132

Kavitha R, Tan TC, Lee HL, Nazni WA, Sofian AM (2013a) Molecular identification of Malaysian Chrysomya megacephala (Fabricius) and Chrysomya rufifacies (Macquart) using life stage specific mitochondrial DNA. Trop Biomed 30(2):211–219

Kavitha R, Tan TC, Lee HL, Nazni WA, Sofian-Azirun M (2013b) DNA typing of Calliphorids collected from human corpses in Malaysia. Trop Biomed 30(1):119–124

Kettle DS (1995) Medical and veterinary entomology, CAB International, 1995, 725 pp

Kim YH, Shin SE, Ham CS, Kim SY, Ko KS, Jo TH, Son GH, Park SH, Hwang JJ (2014) Molecular identification of necrophagous muscidae and sarcophagidae fly species collected in Korea by mitochondrial cytochrome C oxidase subunit I nucleotide sequences. ScientificWorldJournal 2014:1

Kulshrestha P, Chandra H (1987) Time since death. An entomological study on corpses. Am J Forensic Med Pathol 8(3):233–238

Kumara TK, Disney RH, Abu HA et al (2012) Occurrence of oriental flies associated with indoor and outdoor human remains in the tropical climate of North Malaysia. J Vector Ecol 37:62–68

Lessinger AC, Martins Junqueira AC, Lemos TA, Kemper EL, da Silva FR, Vettore AL, Arruda P, Azeredo-Espin AML (2000) The mitochondrial genome of the primary screwworm fly Cochliomyia hominivorax (Diptera: Calliphoridae). Insect Mol Biol 9:521–529

Limsopatham K, Klong-Klaew T, Fufuang N, Sanit S, Sukontason KL, Sukontason K, Somboon P, Sontigun N (2021) Wing morphometrics of medically and forensically important muscid flies (Diptera: Muscidae). Acta Trop 222:106062

Limsopatham K, Tanajitaree C, Sanit S, Sukontason K, Somboon P, Amendt J, Feddern N, Sukontason KL (2022) Wing morphometrics as a tool for the identification of forensic important Lucilia spp. (Diptera: Calliphoridae). Acta Trop 226:106242

Liu Y, Chen Y, Guo Y et al (2013) Estimation of post-mortem interval for a drowning case by using flies (Diptera) in central-South China: implications for forensic entomology. Rom J Leg Med 21:293–298

Lo Pinto S, Giordani G, Tuccia F, Ventura F, Vanin S (2017) First records of Synthesiomyia nudiseta (Diptera: Muscidae) from forensic cases in Italy. Forensic Sci Int 276:e1–e7

Martinson EO, Peyton J, Kelkar YD, Jennings EC, Benoit JB, Werren JH, Denlinger DL (2019) Genome and Ontogenetic-Based Transcriptomic Analyses of the Flesh Fly, Sarcophaga bullata. G3 (Bethesda) 9(5):1313–1320. https://doi.org/10.1534/g3.119.400148

Mashaly AM (2016) Entomofaunal succession patterns on burnt and unburnt rabbit carrion. J Med Entomol 53:296–303

Mashaly A, Alajmi R, Mustafa AE, Rady A, Alkhedir H (2017) Species abundance and identification of forensically important flies of Saudi Arabia by DNA barcoding. J Med Entomol 54(4): 837–843

Matuszewski S, Frątczak K, Konwerski S et al (2016) Effect of body mass and clothing on carrion entomofauna. Int J Legal Med 130:221–232

Meiklejohn KA, Wallman JF, Dowton M (2011) DNA-based identification of forensically important Australian Sarcophagidae (Diptera). Int J Legal Med 125:27–32

Meiklejohn KA, Wallman JF, Dowton M (2013) DNA barcoding identifies all immature life stages of a forensically important flesh fly (Diptera: Sarcophagidae). J Forensic Sci 58:184–187

Michalski M, Gadawski P, Klemm J, Szpila K (2021) New species of soldier Fly-Sargus bipunctatus (Scopoli, 1763) (Diptera: Stratiomyidae), recorded from a human corpse in Europe-a case report. Insects 12(4):302

Monum T, Sukontason KL, Sribanditmongkol P, Sukontason K, Samerjai C, Limsopatham K, Suwannayod S, Klong-Klaew T, Wannasan A (2017) Forensically important blow flies chrysomya *pinguis*, *C. villeneuvi*, and *Lucilia porphyrina* (Diptera: Calliphoridae) in a case of human remains in Thailand. Korean J Parasitol 55(1):71–76

Nandi BC (2004) Checklist of Calliphoridae (Diptera) of India, rec. Zool. Surv. India, dcc. Paper No. 231: 1–47. (Published by the Director, Zoological Survey. India, Kolkata)

Napoleão KS, Mello-Patiu CA, Oliveira-Costa J et al (2016) DNA-based identification of forensically important species of Sarcophagidae (Insecta: Diptera) from Rio de Janeiro, Brazil. Genet Mol Res 15:1–7

Olekšáková T, Žurovcová M, Klimešová V, Barták M, Šuláková H (2018) DNA extraction and barcode identification of development stages of forensically important flies in The Czech Republic. Mitochondrial DNA A DNA Mapp Seq Anal 29(3):427–430

Oliveira TC, Vasconcelos SD (2010) Insects (Diptera) associated with cadavers at the Institute of Legal Medicine in Pernambuco, Brazil: implications for forensic entomology. Forensic Sci Int 198:97–102

Otranto D, Stevens JR (2002) Molecular approaches to the study of myiasis-causing larvae. Int J Parasitol 32(11):1345–1360

Park SH, Zhang Y, Piao H, Yu DH, Jeong HJ, Yoo GY, Chung U, Jo TH, Hwang JJ (2009) Use of cytochrome c oxidase subunit i (COI) nucleotide sequences for identification of the Korean Luciliinae fly species (Diptera: Calliphoridae) in forensic investigations. J Korean Med Sci 24(6):1058–1063

Park SH, Shin SE (2013) Molecular species identification of forensically important flies in Korea. Korean J Leg Med 37(4):177–182

Park JH, Kim YH, Ham CS, Shin SE, Lee HJ, Ko KS, Choi J, Son GH, Park SH (2018) Molecular identification of forensically important calliphoridae and sarcophagidae species using ITS2 nucleotide sequences. Forensic Sci Int 284:1–4

Pastula EC, Merritt RW (2013) Insect arrival pattern and succession on buried carrion in Michigan. J Med Entomol 50:432–439

Picard CJ, Johnston JS, Tarone AM (2012) Genome sizes of forensically relevant Diptera. J Med Entomol 49(1):192–197

Pinto J, Magni PA, O'Brien RC, Dadour IR (2021) Forensically relevant blow flies (Diptera: Calliphoridae) of Central Connecticut, USA. Forensic Sci Int 327:110940

Pohjoismäki JL, Karhunen PJ, Goebeler S et al (2010) Indoors forensic entomology: colonization of human remains in closed environments by specific species of sarcosaprophagous flies. Forensic Sci Int 199:38–42

Povolny D, Verves YG (1997) The flesh-flies of Central Europe (Insecta, Diptera, Sarcophagidae). Spixiana Suppl 24:1–260

Qiu D, Cook CE, Yue Q, Hu J, Wei X, Chen J, Liu D, Wu K (2017) Species-level identification of the blowfly Chrysomya megacephala and other Diptera in China by DNA barcoding. Genome 60(2):158–168

Ratnasingham S, Hebert PD (2007) Bold: the barcode of life data system (http://www.barcodinglife.org). Mol Ecol Notes 7(3):355–364

Reibe S, Schmitz J, Madea B (2009) Molecular identification of forensically important blowfly species (Diptera: Calliphoridae) from Germany. Parasitol Res 106(1):257–261

Ren L, Shang Y, Chen W, Meng F, Cai J, Zhu G, Chen L, Wang Y, Deng J, Guo Y (2018) A brief review of forensically important flesh flies (Diptera: Sarcophagidae). Forensic Sci Res 3(1): 16–26

Rognes K (1992) Revision of the cluster-flies of the Pollenia venturii species-group with a cladistic analysis of Palaearctic species of Pollenia Robineau-Desvoidy (Diptera: Calliphoridae). Insect Syst Evol 23:233–248

Rolo EA, Oliveira AR, Dourado CG, Farinha A, Rebelo MT, Dias D (2013) Identification of sarcosaprophagous Diptera species through DNA barcoding in wildlife forensics. Forensic Sci Int 228(1–3):160–164

Roziah A, Tan SH, Lee HL et al (2015) Mitochondrial and nuclear DNA for identification of forensically important flesh flies (Sarcophagidae: Boettcherisca Spp). Entomol Ornithol Herpetol 4:163

Salem AM, Adham FK, Picard CJ (2015) Survey of the genetic diversity of forensically important Chrysomya (Diptera: Calliphoridae) from Egypt. J Med Entomol 52(3):320–328

Schroeder H, Klotzbach H, Elias S, Augustin C, Pueschel K (2003) Use of PCR-RFLP for differentiation of calliphorid larvae (Diptera, Calliphoridae) on human corpses. Forensic Sci Int 132(1):76–81

Sharma M, Singh D, Sharma AK (2015) Mitochondrial DNA based identification of forensically important Indian flesh flies (Diptera: Sarcophagidae). Forensic Sci Int 247:1–6

Sharma A, Drijfhout FP, Tomberlin JK, Bala M (2021a) Cuticular hydrocarbons as a tool for determining the age of Chrysomya rufifacies (Diptera: Calliphoridae) larvae. J Forensic Sci 66(1):236–244

Sharma A, Tomberlin JK, Delclos P, Bala M (2021b) Volatile compounds reveal age: a study of volatile organic compounds released by Chrysomya rufifacies immatures. Int J Legal Med 135(3):967–977

Silahuddin SA, Latif B, Kurahashi H et al (2015) The importance of habitat in the ecology of decomposition on rabbit carcasses in Malaysia: implications in forensic entomology. J Med Entomol 52:9–23

Singh D, Bala M (2009) The effect of starvation on the larval behavior of two forensically important species of blow flies (Diptera: Calliphoridae). Forensic Sci Int 193(1–3):118–121

Singh D, Bharti M (2001) Further observations on the nocturnal oviposition behaviour of blow flies (Diptera: Calliphoridae). Forensic Sci Int 120(1–2):124–126

Singh D, Bharti M (2008) Some notes on the nocturnal larviposition by two species of Sarcophaga (Diptera: Sarcophagidae). Forensic Sci Int 177:19–20

Singh D, Garg R, Wadhawan B (2012) Ultramorphological characteristics of immature stages of a forensically important fly Parasarcophaga ruficornis (Fabricius) (Diptera: Sarcophagidae). Parasitol Res 110(2):821–831

Song Z, Wang X, Liang G (2008) Species identification of some common necrophagous flies in Guangdong province, southern China based on the rDNA internal transcribed spacer 2 (ITS2). Forensic Sci Int 175:17–22

Sontigun N, Sukontason KL, Zajac BK, Zehner R, Sukontason K, Wannasan A, Amendt J (2017) Wing morphometrics as a tool in species identification of forensically important blow flies of Thailand. Parasit Vectors 10(1):229

Sperling FA, Anderson GS, Hickey DA (1994) A DNA-based approach to the identification of insect species used for postmortem interval estimation. J Forensic Sci 39(2):418–427

Stevens J, Wall R (2001) Genetic relationships between blowflies (Calliphoridae) of forensic importance. Forensic Sci Int 120(1–2):116–123

Stevens JR, Wall R, Wells JD (2002) Paraphyly in Hawaiian hybrid blowfly populationsand the evolutionary history of anthropophilic species. Insect Mol Biol 11:141–148

Sukontason K, Bunchu N, Chaiwong T et al (2010) Forensically important flesh fly species in Thailand: morphology and developmental rate. Parasitol Res 106:1055–1064

Syamsa RA, Ahmad FM, Marwi MA et al (2010) An analysis of forensic entomological specimens by Universiti Kebangsaan Malaysia. Med J Malaysia 65:192–195

Syamsa RA, Omar B, Zuha RM et al (2015) Forensic entomology of high-rise buildings in Malaysia: three case reports. Trop Biomed 32:291

Syamsa RA, Omar B, Ahmad FM et al (2017) Comparative fly species composition on indoor and outdoor forensic cases in Malaysia. J Forensic Legal Med 45:41–46

Szpila K, Mądra A, Jarmusz M et al (2015) Flesh flies (Diptera: Sarcophagidae) colonising large carcasses in Central Europe. Parasitol Res 114:2341–2348

Szpila K, Żmuda A, Akbarzadeh K, Tofilski A (2019) Wing measurement can be used to identify European blow flies (Diptera: Calliphoridae) of forensic importance. Forensic Sci Int 296:1–8

Tan SH, Rizman-Idid M, Mohd-Aris E, Kurahashi H, Mohamed Z (2010) DNA-based characterisation and classification of forensically important flesh flies (Diptera: Sarcophagidae) in Malaysia. Forensic Sci Int 199(1–3):43–49. https://doi.org/10.1016/j.forsciint.2010.02.034

Thongjued K, Chotigeat W, Bumrungsri S, Thanakiatkrai P, Kitpipit T (2019) A new cost-effective and fast direct PCR protocol for insects based on PBS buffer. Mol Ecol Resour 19(3):691–701

Tomberlin JK, Benbow ME (2015) Forensic entomology international dimensions and frontiers. CRC Press, Boca Raton (FL)

Toukairin Y, Arai T, Hoshi T et al (2017) The geographical distribution of fly larvae on corpses in Saitama prefecture in Japan during the summer season. Leg Med (Tokyo) 24:75–77

Vasconcelos SD, Cruz TM, Salgado RL et al (2013) Dipterans associated with a decomposing animal carcass in a rainforest fragment in Brazil: notes on the early arrival and colonization by necrophagous species. J Insect Sci 13:145

Vasconcelos SD, Soares TF, Costa DL (2014) Multiple colonization of a cadaver by insects in an indoor environment: first record of Fannia trimaculata (Diptera: Fanniidae) and Peckia (Peckia) chrysostoma (Sarcophagidae) as colonizers of a human corpse. Int J Legal Med 128:229–233

Velásquez Y, Magaña C, Martínez-Sánchez A et al (2010) Diptera of forensic importance in the Iberian Peninsula: larval identification key. Med Vet Entomol 24:293–308

Voss SC, Magni P, Dadour I, Nansen C (2017) Reflectance-based determination of age and species of blowfly puparia. Int J Legal Med 131(1):263–274

Wallman JF, Donnellan SC (2001) The utility of mitochondrial DNA sequences for the identification of forensically important blowflies (Diptera: Calliphoridae) in southeastern Australia. Forensic Sci Int 120(1–2):60–67

Wang J, Li Z, Chen Y et al (2008) The succession and development of insects on pig carcasses and their significances in estimating PMI in South China. Forensic Sci Int 179:11–18

Wang Y, Ma MY, Jiang XY et al (2017) Insect succession on remains of human and animals in Shenzhen. China Forensic Sci Int 271:75–86

Wells JD, Smith JL (2013) First report of *Blaesoxipha plinthopyga* (Diptera: Sarcophagidae) from a human corpse in the U.S.A. and a new state geographic record based on specimen genotype. J Forensic Sci 58:1378–1380

Wells JD, Stevens JR (2008) Application of DNA-based methods in forensic entomology. Annu Rev Entomol 53:103–120

Wells JD, Pape T, Sperling FA (2001) DNA-based identification and molecular systematics of forensically important Sarcophagidae (Diptera). J Forensic Sci 46:1098–1102

Ying BW, Liu TT, Fan H, Wei D, Wen FQ, Bai P, Huang J, Hou YP (2007) The application of mitochondrial DNA cytochrome oxidase II gene for the identification of forensically important blowflies in Western China. Am J Forensic Med Pathol 28(4):308–313

Yusseff-Vanegas S, Agnarsson I (2016) Molecular phylogeny of the forensically important genus Cochliomyia (Diptera: Calliphoridae). Zookeys (609):107–120

Zaidi F, Wei SJ, Shi M, Chen XX (2011) Utility of multi-gene loci for forensic species diagnosis of blowflies. J Insect Sci 11:59

Zajac BK, Sontigun N, Wannasan A et al (2016) Application of DNA barcoding for identifying forensically relevant Diptera from northern Thailand. Parasitol Res 115(6):2307–2320

Zehner R, Amendt J, Schutt S et al (2004) Genetic identification of forensically important flesh flies (Diptera: Sarcophagidae). Int J Legal Med 118:245–247

Zhang C, Fu X, Xie K et al (2015a) MtDNA analysis for genetic identification of forensically important Sarcophagid flies (Diptera: Sarcophagidae) in China. J Med Entomol 52:1225–1233

Zhang CQ, Fu XL, Yang X et al (2015b) Application of MtSNP marker for genetic identification of forensically important Sarcophagid flies (Diptera: Sarcophagidae) in China. Forensic Sci Int Gen Suppl 5:240–242

Zumpt F (1965) Myiasis in man and animals in the old world, vol 15. Butterworths, London, pp 1–267

Forensic Entomology and the Veterinary World

7

Abstract

Although our knowledge on the forensic entomology has improved over the years, more developments have happened encompassing the studies from human death. Forensic entomology is a far recent development but is progressing by the day. We discuss in this chapter some of the major developments and different aspects in the fields of forensic entomology in the veterinary world and a few studies. It appears that animal cruelty in both domestic and wild animals in veterinary forensics are comparatively neglected area that needs far more focus of research and development.

Keywords

Forensic entomology · Veterinary world

Abbreviations

C loewi	*Calliphora loewi*
C macellaria	*Cochliomyia macellaria*
C uralensis	*Calliphora uralensis*
C vicina	*Calliphora vicina*
C vomitoria	*Calliphora vomitoria*
COI	Cytochrome c oxidase I
D maculatus	*Dermestes maculatus*
H anxia	*Hydrotaea anxia*
L caesar	*Lucilia caesar*
L cuprina	*Lucilia cuprina*
L sericata	*Lucilia sericata*
M domestica	*Musca domestica*

S. Ghosh, D. Banerjee, *Biology of Forensically Important Invertebrates*,
https://doi.org/10.1007/978-981-99-5026-3_7

N vespillo	*Nicrophorus vespillo*
N vespilloides	*Nicrophorus vespilloides*
P terraenovae	*Protophormia terraenovae*
P terraenovae	*Protophormia terraenovae*
PMI	Post mortem interval
PMImin	Minimum postmortem interval
S dux	*Sarcophaga dux*
T sinuatus	*Thanatophilus sinuatus*
TOD	Time of death

7.1 Introduction

Forensic entomology is the science of applying necrophagous insects inhabiting dead, decomposed remains for legal investigations. Although it finds significant importance in detecting human criminal cases, forensic entomology is also vital for practicing veterinarians in order to detect veterolegal situations of animal cruelty, like cases of neglect, use of toxin/poison, intentional or malicious nature of abuse, etc., death by illegal poaching, or illegally killed wild animals.

Forensic case of animal death involves both domestic and wild animals. Veterinarians are professionals who primarily investigate animal cruelty cases, examine them, and become involved in a legal investigation. They document cases of abuse, record evidence of poisoning, collect associated physical evidence, and give opinions as experts in a court of law.

Sometimes such investigations involve expert opinion and technical advice from a pathologist, toxicologist, and other specialists.

The everyday veterolegal situations faced by veterinarians include (1) wound examination, (2) poisonings, and (3) postmortem examination.

The science of veterinary forensics is a multidisciplinary subject that encompasses diverse streams of (a) law, (b) toxicology, (c) veterinary medicine, (d) entomology, (e) anthropology, (f) pathology, (g) ballistics, (h) botany, (i) odontology, (j) DNA analysis, and (k) other forensic analyses.

We discuss in this chapter, in brief, the (1) vet and veterolegal cases.

7.2 Veterinary and Veterolegal Cases

Veterinary forensic laboratory functions under a team of forensic veterinary professionals, and their expertise are used in veterolegal cases involving cases of:

(a) Poaching
(b) Traffic accidents of animals
(c) Animal deaths in a zoo

(d) Farm animal deaths
(e) Suspected cruelty to animals

Death investigations involve finding the cause of death, including a complete necropsy.

Investigations include finding of (1) the death cause, (2) toxicology, (3) analysis of DNA, (4) identification of species, (5) examination of morphology, (6) animal welfare, (7) crime scene, (8) ballistics, (9) macroscopical examination and histology, (10) detection of bite injuries by DNA testing, (11) paternity testing in costly species like in horse breeding, (12) food safety examinations involving DNA analysis, in case of suspected food contaminants, (13) detection of the bite mark, scratch mark, and trauma (14) wound analysis, and (15) bone analysis.

Other situations handled by a forensic vet include:

1. Cases of suspected cruelty
2. Neglect of farm and pet animals
3. Inappropriate animal shelter conditions
4. Non-lethal injuries to animals like events of the shooting, stabbing, beating, and hunting by illegal methods using snares and illegal weapons
5. Poisoning, which may be intentional or accidental

Although awareness of animal cruelty, the legal cause of death of animals, and the role of forensic vets is growing among livestock owners and the public at large, in some cases, it is difficult to conclude the crime. It requires an integrated knowledge from a veterinary forensic sciences laboratory, together with inputs from veterinary pathology, medicine, toxicology, forensic entomology, animal biotechnology, and other allied sciences.

A veterinarian is well-equipped with the knowledge of animal behavior and clinical experience to evaluate injuries. The injury type and fatal injury in deceased victims need to be determined, and then the crime incident can be backtracked. It is often challenging to locate the site of hemorrhage and multiple fractures or injuries, which need to be evaluated like those in the skull, where a single blow may generate concentric or radiating fractures. The animal's response to the injury also needs to be evaluated.

In case of poisoning in livestock due to personal enmity among animal owners, it is vital to understand intentional poisoning, done by using agents like anticoagulant rodenticides, carbamates, cyanide, organophosphates, organochlorine, strychnine, and thallium to accidental exposure of animals by pesticides.

In order to determine the time of death (TOD) (Table 7.1), postmortem interval (PMI) needs to be determined. Environmental conditions and the condition of the body are to be recorded.

Table 7.1 TOD estimates in animals

Condition of dead body	Time (h) elapsed
Body warm but not stiff	0–3 h
Body warm and stiff	3–8 h
Body cold and stiff	8–36 h
Body cold but not stiff	36 h

7.3 Forensic Entomology for Animal Deaths

Analysis on insects and their life cycle stages finds application in determining TOD. Maggots enable determination of location and TOD through PMI and provide DNA and toxicological evidence.

Flies lay eggs under certain environmental conditions, at certain times of day after death of an animal and this is species specific. These eggs then hatch into maggots which are influenced by environmental conditions. The developmental rate of the larvae develop is species specific and based on environmental conditions.

Blow flies are the first insects to be attracted to the dead body. So by estimating the time of colonization or laying of eggs by blow flies, the TOD can be estimated. Other forensically important insects like beetles feed at different times on dead body.

There are several ways used to document like:

1. Photo documentation for forensic necropsies, with significant findings and support to identify the specimen.
2. Specimen collection is another very important method used to determine PMI and cause of death.

   During the necropsy, a pathologist needs to collect the proper samples in an appropriate manner.
3. In suspected cases of poisoning, stomach and intestine upper part and its contents and a part of the liver, kidney, and spleen need to be collected for testing. Evidences collected, like projectiles/tissue samples, need to be labeled and documented prior to examination.
4. Tissue samples need to be collected for histopathological examination or testing and detection of the disease.
5. Small piece of tissues and the lesion needs to be collected in a glass container with 10% formalin and labeled appropriately and packed carefully before dispatched to labs for testing.

Vets can get involved in:

1. Medicolegal proceedings by assisting authorities in legal proceedings
2. Animal deaths due to road accidents, certifying the nature of injuries and the cause of death and acts as a government witness
3. In cases of bestiality, examination of vaginal washing for presence of human sperms, and based on which, the court decides the punishment to the person

4. Detecting frauds in animals done with animals to alter their identity
5. In case of theft, brought by police, for proper identity of the animal in order to be handed over to the right owner and accord punishment to the culprit
6. Detection of adulteration and animal products like meat and meat products, milk, ghee, wool, and bristles important from the point of view of quality control and penalizing the fraudulent dealer
7. Preventing illegal cow slaughter where it is banned
8. In cases related to export and import of animal and animal products

Insect analysis is used in determining TOD. Maggots help in determining the location and time of death through PMI and provide DNA and toxicology evidence. Flies lay eggs under certain environmental conditions, at certain times of day after the death of an animal, and are species-specific.

They hatch into maggots and hatching of eggs are influenced by environmental conditions. The developmental rate of larvae is species-specific and based on environmental conditions.

Blow flies are the first to be attracted to the dead body, so the TOD can be estimated by dating the time of colonization or laying of eggs. Other forensically important insects like beetles feed at different times on dead bodies.

7.4 Forensic Entomology and Veterinary World

Forensic entomology works on determining minimum postmortem interval (PMImin) by identifying the arthropods and insects colonizing the dead body and passing a life cycle on the dead body or animal/human carcass and thereby estimating its colonization time, based on the prior knowledge of the developmental rate of the colonizing arthropod insect species and succession in the decomposed dead body.

The TOD has been associated with temperature and other weather conditions at the crime site since arthropods, including insect development on the corpse, depending on temperature and weather conditions. Thus, accurate recording of weather conditions at the crime site and where the dead body is remains to find necessary.

These arthropods, including insect fly larvae, are essential decomposers of both primary and secondary order that decompose the animal remains.

It is possible quite accurately to estimate the time of death based on the knowledge of the developmental biology of insect decomposer species at different temperatures.

The pattern of succession of arthropods and developmental stages of different arthropod species found on decomposed cadavers highlight death investigations and provide an estimate of the time elapsed since death.

Although the application of forensic entomology at crime scenes to solve crimes dates back centuries, the majority of studies exist on human corpses. Bibliometric analysis of forensic entomology through the last two decades reveals that the order

Diptera and the family Calliphoridae form the most studied groups, and the PMI estimation and species identification were the most common topics (Lei et al. 2019). Its use in legal, veterinary medicine, and forensic entomology cases involving animals is new and has been studied recently. Information from the literature available on human death or abuse investigation finds importance in understanding the veterinary field, but the field is developing (Crook 2007; Huffman and Wallace 2012; Anderson et al. 2001; Benecke 2001; Williams and Villet 2006; Defilippo et al. 2016; Amendt et al. 2009; Anderson 1995; Bugelli et al. 2015; Byrd and Castner 2010; Dadour et al. 2001).

Forensic entomology is important in investigating cases of perimortem and postmortem treatment of remains. It demonstrates human and animal neglect, enabling understanding of the relation between a suspect, a victim, and events in a crime scene and can aid in detecting drugs/toxins within decomposed or skeletonized remains (Benecke et al. 2004; Campobasso et al. 2005; Introna et al. 2001; Brundage and Byrd 2016; Forensic Working Group 2014).

Insects under two significant orders involved in forensic entomology include:

1. Blow flies including flesh flies, house flies, and cheese skippers under Order Diptera
2. Beetles, including carrion beetles, dermestids, and scarab beetles under Order Coleoptera (Byrd and Castner 2001)

The insect activity on the dead body that is undergoing decomposition leads to release of odor (Table 7.2).

Post-death, the decay process releases odors that attract insects to the decomposing carcass releasing volatile organic compounds (VOCs) (Verheggen et al. 2017). The flies under Order Diptera, including blow flies under Family Calliphoridae and Muscidae and flesh flies under family Sarcophagidae and beetles, arrive on the decomposing carcass, feeding on flesh, facilitating the decay process. Decomposing enzymes, sulfur, and methane compounds attract insects and sometimes act as stimulants for oviposition (Table 7.3).

In the early stages of decomposition, competition exists among colonizers as a cadaver is rich in resources. Diverse arthropod groups are adapted to different stages of decomposition, including species with a preference for feeding on a fresh carcass and those with a preference for feeding on the dead, decayed matter of 2 weeks old. The events of decay and forensic fly activity on decaying dead body are highlighted in Table 7.4.

A study on the insect succession on the dead body reveals that the blow flies are the first among other flies that are attracted by the dead body and they inhabit, feed, and start a life cycle on the dead body.

The insect succession has been studied on dead decaying bodies revealing a gradual colonization by the flies at various stages and time of death. Smith in 1986 proposed a model for insect succession on the dead decayed bodies (Fig. 7.1).

An entomologist has multifarious roles in the crime scene detecting crime based on study of the insects, life cycle stages, their identification and characterization,

Table 7.2 Vertebrate decomposition in outdoor terrestrial environments and relevant factors during each stage

	Stage	Description	Insect activity	Odor evolution
Early decomposition	Fresh	First hours following death	Initial attraction of flies to carcass and oviposition in body orifices	Minimal perceived odor
		Onset of autolysis		
		Rigor mortis, livor mortis, algor mortis		
	Bloat	Onset of putrefaction	Continued attraction and oviposition	Strong perceived odor
		Gas accumulation due to microorganisms and macromolecule breakdown	Development of fly larvae and initial larval masses formed	Odor typically perceived as containing sulfur compounds, methane, and ammonia
Late decomposition	Active decay	Deflation of the body	Active consumption of tissue by matured insect larvae	Strong and complex odor comprising a wide array of compounds
		High rate of tissue removal by microorganisms, insect activity, and macromolecule breakdown		
	Advanced decay	Reduction in moisture	Reduced insect activity and migration of fly larvae	Reduction in perceived odor but still complex and present
		Some remaining tissue typically in torso region	Colonization by beetles	
	Dry/ remains	Hair, cartilage and bone remain	Minimal fly and beetle activity	Reduced odor
		Mineral degradation	Regular assemblage re-established	Perceived odor mostly from surrounding soil

Table adapted with permission from Verheggen et al. (2017)

geography, study of the location, and collecting data on the environmental condition (Table 7.5) and use relevant information in calculating the PMI with greater accuracy (Table 7.6).

A simple formula used to calculate degree-day accumulation for a species with a 10 °C threshold is as follows:

$$\frac{(\text{DailyMax} - 10) + \text{DailyMax} - 10}{2}$$

If max or min is less than 10, then enter zero (Gujar et al. 2020).

Entomotoxicology finds importance in identifying and detecting toxins in the invertebrate decomposers that are used to highlight the cause of death.

Table 7.3 Table of decomposition VOCs affecting carrion insects[a]

Receiving individual	Species	Cadaveric volatile organic compound (s)	Behavioral response	References
Diptera				
Calliphoridae	*Calliphora vomitoria*	Dimethyl trisulfide	Attractant	Nilssen et al. (1996)
	Calliphora vicina	Heptanal; nonanal; dimethyl mono-, di-, and trisulfide	Attractant	Aak and Knudsen (2011), Johansen et al. (2014), Stensmyr et al. (2002)
		Butylated hydroxyl toluene; 3-hydroxy-2-butanone	Possible attractant	Johansen et al. (2014)
	Calliphora uralensis	Dimethyl trisulfide	Attractant	Nilssen et al. (1996)
	Calliphora loewi	Dimethyl trisulfide	Attractant	Nilssen et al. (1996)
	Cochliomyia macellaria	A mixture of isovaleraldehyde; 4-methylpentan-2-one; dimethyl disulfide; dimethyl trisulfide; 3-methyl butanol; 2-phenylethanol; P-cresol	Oviposition stimulant	Zhu et al. (2013)
	Lucilia sericata	Indole; dimethyl trisulfide	Oviposition stimulant	Brodie et al. (2014), Brodie et al. (2016), Cragg and Ramage (1945), Cragg and Thurston (1950), Frederickx et al. (2012)
		Ethanethiol; dimethyl disulfide; dimethyl trisulfide; butan-1-ol; p-cresol; phenol; indole	Attractant	
	L caesar	Ethanethiol; dimethyl mono-, di-, and trisulfide; hydrogen sulfide; indole	Attractant	Cragg and Ramage (1945), Cragg and Thurston (1950), Stensmyr et al. (2002)
	L cuprina	Indole	Oviposition stimulant	Browne (1965)
	Protophormia terraenovae	Dimethyl trisulfide	Attractant	Nilssen et al. (1996)

(continued)

Table 7.3 (continued)

Receiving individual	Species	Cadaveric volatile organic compound (s)	Behavioral response	References
Muscidae	*M domestica*	Mix of butanoic acid, 3-methylindole and dimethyl trisulfide	Attractant	Cossé and Baker (1996)
	H anxia	Dimethyl trisulfide	Attractant	Nilssen et al. (1996)
Coleoptera				
Silphidae	*Thanatophilus sinuatus*	Dimethyl disulfide	Attractant	Dekeirsschieter et al. (2013)
	Nicrophorus vespillo	Dimethyl sulfide, dimethyl disulfide, dimethyl trisulfide	Attractant	Kalinová et al. (2009), Podskalská et al. (2009)
	Nicrophorus vespilloides	Dimethyl sulfide, dimethyl disulfide, dimethyl trisulfide	Attractant	Kalinová et al. (2009)
Dermestidae	*Dermestes maculatus*	Benzyl butanoate	Attractant	Von Hoermann et al. (2012)
Hymenoptera				
Pteromalidae	*Nasonia vitripennis*	Dimethyl disulfide	Attractant	Frederickx et al. (2014)

[a]Table adapted with permission from Verheggen et al. (2017)

However, studies from across the globe reveal a growing need to increase awareness of forensic entomology among practicing veterinarians. More need for education, research and development, and training in these lines is the need of the hour so that veterinarians are well equipped to practice forensic entomology in saving wildlife (Gujar et al. 2020).

7.5 Forensic Entomology and Crimes on Animals

Forensic entomology is helpful to veterinary professionals in:

1. The detection of cases of animal cruelty, involving the determination of the PMImin interval identification of the forensic insect species and estimating the time of colonization of insect species.
2. This is based on prior knowledge of the insect species and its development rate, life cycle, and on succession of insect species during the decomposition of animal remains.
3. Wound detection and recognition and estimation of the time of periods of neglect.
4. Detecting cases of illegal poaching.
5. Multiple factors require documentation in order to understand the crime.

Table 7.4 The stages of decomposition, putrification, and forensic entomology

Stage	Days after death	Events
Fresh	1–2 days	1. Starts at death and ends with bloating. During this stage, 2. Disintegration of biomacromolecules like carbohydrates and protein into simpler compounds 3. Calliphoridae larvae consume most soft tissue on the remains within a few days
Bloated	2–6 days	1. The petrification process starts 2. Generation and release of gasses generated by anaerobic bacteria 3. Abdominal inflammation 4. Flesh flies appear 5. Blow flies are still in oviposition
Decay	5–11 days	1. Starts at around 5 days 2. Leakage of abdominal walls enables the release of gasses 3. Decomposed fluid seepage from the body, 4. Carcass reduces (~20% of original mass) 5. Adult blow flies are not attracted 6. Maggots feed on a dead body internally and externally 7. Predators and parasites of maggots appear
Post-decay stage	10–25 days	1. Cadaver becomes dry 2. Large maggots leave 3. Mites and other predatory beetle increase 4. Hide beetles predominant in dry condition 5. Skin, cartilage, and bones remain in dry environment 6. Wet, viscous matter in the soil under the remains occur in wet environment
Dry stage	25+ days	1. Remains of the carcass, including skin and bones 2. Insects, including clothes moths and Dermestid beetles, digest keratin 3. Maggots stay long in a wet region 4. Hide beetles does not appear 5. Mass of corpse reduced (~10% of the original mass)

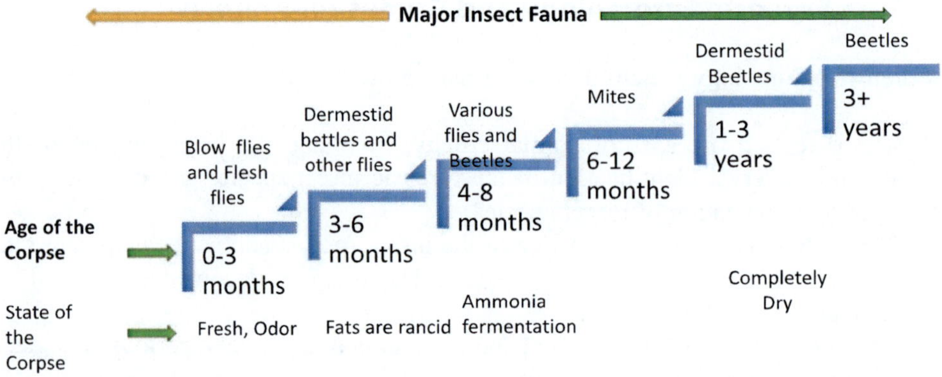

Fig. 7.1 Insect succession on dead body Image adapted from Smith (1986)

Table 7.5 Role of forensic entomologist

(1)	Identify insects and the its life cycle stages of: 1. Eggs 2. Larvae 3. Pupa 4. Adults
(2)	Collection and preservation of insects
	1. Estimating the PMI using factors like: (a) Insect evidence (b) Climate and weather (c) Geography and location (d) Condition of the dead body, etc.
(3)	Evidence in court

Table 7.6 Postmortem interval (PMI)

The following are required to calculate the PMI with accuracy:	
(i)	Collection and identification of the species of insect correctly
(ii)	Lifecycle model of the fly development
(iii)	Calculating the insects age
(iv)	Use of classification key Mt-DNA-based study
(v)	Fly development arranged into stages
(vi)	Fly development stages require time and are temperature dependent
(vii)	Warmer temperatures reveal faster development Cool temperatures reveal slower development
(viii)	The decomposition process highly depends on degree day accumulation

These include:

1. Climatic factors like temperature as the development of insects are temperature dependent.
2. Their geographic distribution, which finds importance as it can highlight any movement of the carcass or suggest associations between the potential suspect, victim, and the crime scene.
3. Drugs, poison, or toxin detection within decomposed remains.
4. Investigation of cases of animal cruelty need documentation and collection and preservation of insect specimens from live animals or during the necropsy (Brundage and Byrd 2016).

Despite the often limited availability of the crime scene and local temperature data, the interpretation of the PMImin and the pathology report have proved vital in detecting animal crimes (McGarry et al. 2018).

Identification of insects at the species level on dead bodies of animals remains a crucial primary task in forensic entomology. This is, again, a challenging task, and morphological determination and detection of immature forms in the life cycle sometimes pose challenges. Thus, in recent times, molecular identification has gained relevance. Cytochrome c oxidase I (COI) gene-based DNA barcoding is advantageous over morphological identification-based studies.

COI-based identification has identified several specimens of seven species of two families of Diptera, including Calliphoridae and Muscidae, found in wildlife carcasses from Portugal (Rolo et al. 2013). Veterinarians in British Columbia reported myiasis and 5–30 cases/per year, and studies revealed the advantages and disadvantages of using forensic entomology in living animals (Anderson and Huitson 2004).

7.6 Forensic Entomology: Animal Deaths—Case Studies

Insects with death sensor mechanism, being the first species to sense death and inhabit the dead organism, aid in predicting postmortem interval (PMI), which finds application in solving human and wildlife crime when sometimes other evidences of crime and prediction of PMI are weak. Forensically important fly species including Sarcophaga (Boettcherisca) peregrina (Robineau-Desvoidy), Sarcophaga (Liopygia) ruficornis (Fabricius), Sarcophaga (Liosarcophaga) dux Thomson, *Sarcophaga albiceps (Meigen), Sarcophaga caerulescens (Zetterstedt), Sarcophaga impatiens (Walker)* have been reported from Australia, China, Egypt, Germany, India, Japan, Malaysia, Singapore, Switzerland and Thailand (Sukontason et al. 2010; Aly and Wen 2013; Aly et al. 2013; Guo et al. 2011; Meiklejohn et al. 2011, 2013). In a study from India, a complete life cycle of *Sarcophaga (Liosarcophaga) dux* Thomson (1869), under Class Insecta, Order Diptera, family Sarcophagidae and subfamily Sarcophaginae has been reported from the yellow-bellied house gecko carrion, *Hemidactylus flaviviridis Ruppel (1835*), growing at around 38 °*C*, relative humidity (RH) of 18–53%, and wind velocity of 21 ± 8 km/h and temperature in the month of May from Kolkata, India, wherein larval to pupa development time was 200 ± 2 h with a total life cycle completion for ~312 h. A total of six species of dipterans were reported on the carcass. The growth and development of *Sarcophaga (L) dux* including first instar larvae, to second and third instar larvae, pupae, imago, and finally emergence of the adult fly (Figs. 7.2, 7.3, and 7.4; Table 7.7) on dead *Hemidactylus flaviviridis Ruppel (1835)* was reported as follows.

7.7 Environmental Forensic Toxicology Legal Perspective

Forensic entomology practitioners face several challenges during presenting their findings to legal professionals, related to terminology of (1) insect age, (2) colonization time, (3) and the limitations of the estimate.

Legal requirements are largely variable in different countries.

Therefore, neither there are any standard format for the entomological case report nor are there any existing guidelines regarding the essential, optional, desirable sections, or unnecessary sections in a entomological report in forensic cases. Several

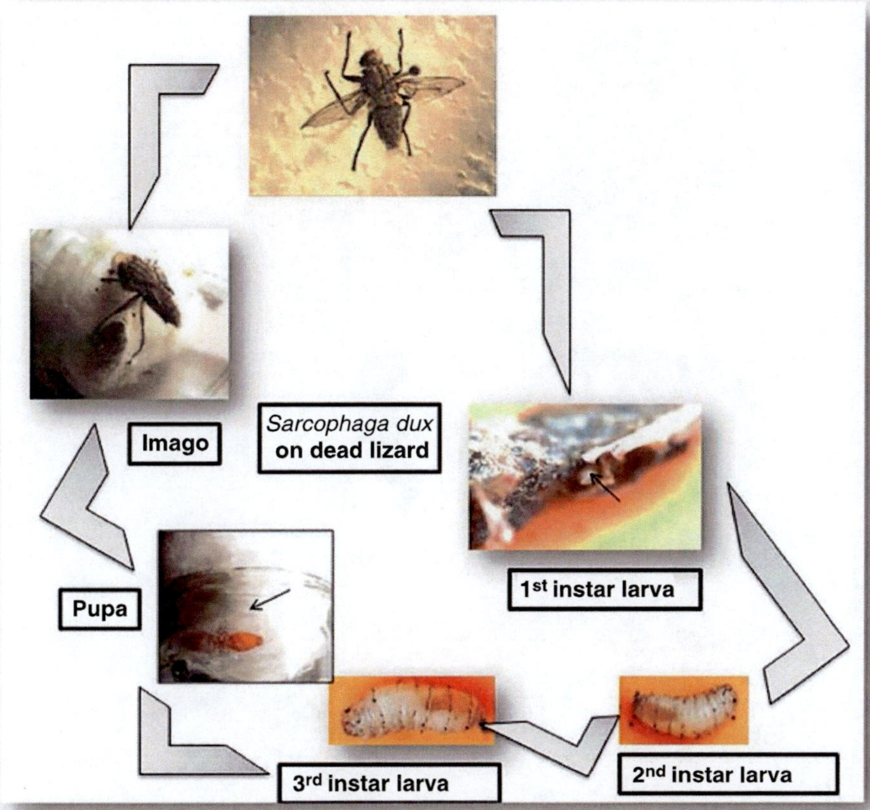

Fig. 7.2 Life cycle of *Sarcophaga dux*, from adult, different instars, pupae, and imago. Figure adapted with permission from Chakraborty et al. (2014)

sections have been proposed to be considered when drafting an entomological case report with criteria to be considered and their limitations.

A global format for presentation of entomological case report seems to be distant as there are differences in national legislation across nations, and enlisting all differences between nations is yet to be established (Kotzé et al. 2021).

The legal investigations in forensic sciences involving domesticated and wild animals received less attention as compared to human forensic. A study of the effect of the fur coat on the activity and development of microbial decomposers in rabbit carcass revealed temporal alterations of metabolic activity and taxa distribution with decomposition process, but the richness and the diversity of the bacterial communities revealed no variation with time and that the carcass fur did not affect either the bacterial communities' functional activity nor the diversity (Tuccia et al. 2019).

Some beetles, like African beetles, with overlapping ecology and geography including *Thanatophilus micans* (Fabricius) and *Thanatophilus mutilatus*

Fig. 7.3 Adult male of *Sarcophaga dux* morphology. Figure adapted with permission from Chakraborty et al. (2014). (**a**) Adult male *S dux*, dorsal view with head, thorax, a part of the abdomen and legs. (**b**) Head and mouth parts of the male *(anterior view)*: the two prominent ommatidia with antennae and frons, sucking mouth parts are shown in the figure.

(Castelneau) that colonize soon after death, are as helpful as blow flies for estimating PMImin. DNA-based identification of these species, their history, and heat summation models for the development of each species have been developed (Ridgeway et al. 2014).

7.8 Discussions

With increasing crime against animals globally, and escalation in cases of poisoning, poaching, and adulteration of meat with other species, veterinary forensics appears to be a much-needed discipline.

Today, with the advances of technology, investigation of animal deaths due to cruelty, mischief, sport, species identification, and crime scene investigation is much easier.

In some countries like India, the lack of national database of animal cruelty statistics makes it difficult to determine cases of crimes committed annually. Environmental Investigation Agency, India, recorded 784 cases of tiger, otter, and leopard skin from poaching between 1994 and 2003 and seizing of 698 otters. This calls in for development of database of animal cruelty in both domestic and wild animals in veterinary forensics.

Fig. 7.3 (continued) Abbreviations used (*po* post ocellar, *oc* ocellar, *vte* outer ventricle, *fr* interfrontal, *forb* fronto-orbital, *vb* vibrissae, *sp v* supra vibrissal, *gn* genal, *st v* sub vibrissal. (**c**) Abdomen (adult male): variable abdominal markings visible. (**d**) Antenna (adult male) Arista plumose on basal second/third, sternopleurals 1:1:1. (**e-f**) Adult male fifth sternite and genital capsule: the V-shaped fifth sternite of the adult male. Bristle-like projections coming out of the fifth sternite, identifying character of *S. dux*. (**e**). Genital capsule along the fifth sternite. Praeputium with upper prong of apical fork longer and stronger than the lower fork. Surstylus, distiphallus, basiphallus. (**f**) *P* phallus, *C* cercus, *Ep* epandrium, *Po* postgonite. *G* Adult male genitalia (side view): the genitalia of an adult male was dissected and viewed under LM. *c* cercus, *ep* epandrium, *prg* pregonite, *j* juxta, *v* vesica, *s* surstylus, *pht* phallic tube. *H* Right fore leg (adult male): the fore leg was dissected and viewed under LM. *c* coxa, *tro* trochanter, *fe* femur, *tib* tibia, *bt* basitarsus or first tarsomer, *tar* tarsomer (2,3,4,5), *pul* pulvillus, *cl* claw. *I* Left fore wing (adult male): the fore wing was visualized under LM and different parts were noted. *c* costal cell, *sc* sub costal, *Sc* sub-costal vein, *C* costal vein, R_1 anterior branch of radius, R_{2+3} radial vein, R_{4+5} radial vein 4 + 5, *M* media, *r-m* radial-medial, *bm-cu* basal-medial-cubital, CuA_1 anterior cubital 1, CuA_2 anterior cubital 2, $A_1 + CuA_2$ first anal vein+ anterior cubital 2

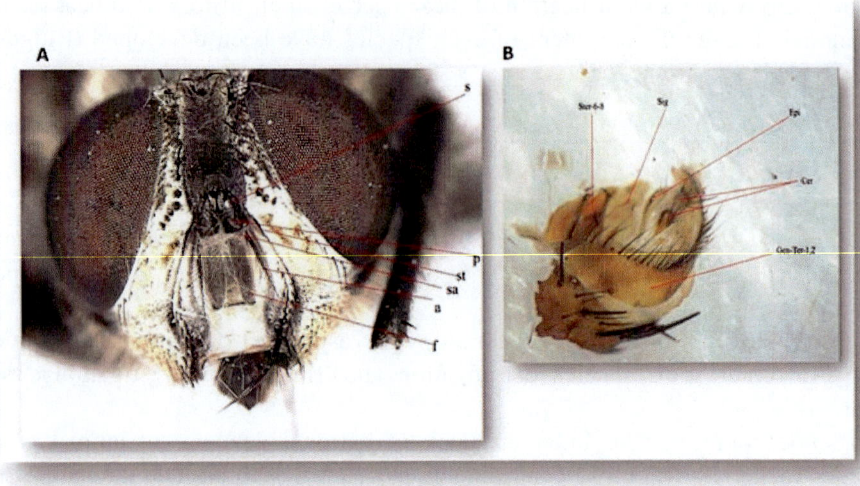

Fig. 7.4 Adult female of *Sarcophaga (Liosarcophaga) dux* morphology. Figure adapted with permission from Chakraborty et al. (2014). (**a**) Front and anterior view of the female *S. dux* head and mouth parts: the front view of the head of an adult female was visualized by LM. *f* first flagellomere, *a* arista, *s* scape, *p* pedicel, *st* bristle, *sa* antennal seam. (**b**) Female genitalia: the female genitalia was dissected out and processed as mentioned in materials and methods. Sternites 6, 7, and 8, signum, epiproct, cerci, and genital tergites 1 and 2 are shown in the figure. *Ster-6-8* sternites 6–8, *Sig* signum, *Epi* epiproct, *Cer* cerci, *Gen-Ter-1, 2* genital tergites 1–2

Table 7.7 Meteorological data and development of *Sarcophaga dux* on dead lizard[a]

No. of days	Time (h)	Average temperature (°C)	Average humidity %	Precipitation (mm of Hg)	Wind speed (Km/h)	Developmental stages
1	24	38 ± 2	44 ± 1	0	29 ± 0	No instar larvae seen
2	48	36 ± 1	49 ± 2	0	27 ± 1	First instar larvae: 26 ± 1 h
3	72	37 ± 1	45 ± 1	0	22 ± 1	Second instar larvae: 56 ± 1 h
4	96	39 ± 2	39 ± 2	0	13 ± 2	Third instar larvae: 156 ± 2 h
5	120	38 ± 2	37 ± 1	0	24 ± 3	
6	144	36 ± 1	32 ± 2	0	21 ± 1	
7	168	38 ± 1	26 ± 3	0	19 ± 2	
8	192	38 ± 2	27 ± 2	0	19 ± 1	Pupae: 200 ± 2 h Imago: 307 ± 3 h Adult: 312 ± 3 h
9	216	39 ± 3	30 ± 1	0	19 ± 2	
10	240	40 ± 1	26 ± 1	0	14 ± 3	
11	264	40 ± 1	28 ± 1	0	13 ± 0	
12	288	39 ± 2	19 ± 1	0	18 ± 1	
13	312	38 ± 1	25 ± 2	0	22 ± 2	

[a]*S.dux* at 38 ± 3 °C took ~13 days to complete its life cycle. *Table adapted from* Chakraborty et al. (2014) *with permission*

References

Aak A, Knudsen GK (2011) Sex differences in olfaction-mediated visual acuity in blowflies and its consequences for gender-specific trapping. Entomol Exp Appl 139:25–34

Aly SM, Wen J (2013) Applicability of partial characterization of cytochrome oxidase I in identification of forensically important flies (Diptera) from China and Egypt. Parasitol Res 112(7):2667–2674

Aly SM, Wen J, Wang X (2013) Identification of forensically important Sarcophagidae (Diptera) based on partial mitochondrial cytochrome oxidase I and II genes. Am J Forensic Med Pathol 34(2):159–163

Amendt J, Zehner R, Johnson DG, Wells J (2009) Future trends in forensic entomology. In: Current concepts in forensic entomology. Springer, Dordrecht, pp 353–368

Anderson GS (1995) The use of insects in death investigations: an analysis of cases in British Columbia over a five-year period. Can Soc Forensic Sc 28(4):277–292

Anderson GS, Huitson NR (2004) Myiasis in pet animals in British Columbia: the potential of forensic entomology for determining duration of possible neglect. Can Vet J 45(12):993–998

Anderson GS, Byrd JH, Castner JL (2001) Insect succession on carrion and its relationship to determining time of death. Forensic entomology: the utility of arthropods in legal investigations, vol 143. CRC Press, Boca Raton, FL, p 76

Benecke M (2001) A brief history of forensic entomology. Forensic Sci Int 120(1–2):2–14

Benecke M, Josephi E, Zweihoff R (2004) Neglect of the elderly: forensic entomology cases and considerations. Forensic Sci Int 146(195):9

Brodie B, Gries R, Martins A, Vanlaerhoven S, Gries G (2014) Bimodal cue complex signifies suitable oviposition sites to gravid females of the common green bottle fly. Entomol Exp Appl 153:114–127

Brodie BS, Babcock T, Gries R, Benn A, Gries G (2016) Acquired smell? Mature females of the common green bottle fly shift semiochemical preferences from feces feeding sites to carrion oviposition sites. J Chem Ecol 42:40–50

Browne BL (1965) An analysis of the ovipositional responses of the blowfly Lucilia cuprina to ammonium carbonate and indole. J Insect Physiol 11:1131–1143

Brundage A, Byrd JH (2016) Forensic entomology in animal cruelty cases. Vet Pathol 53(5): 898–909

Bugelli V, Forni D, Bassi LA, Di Paolo M, Marra D, Lenzi S et al (2015) Forensic entomology and the estimation of the minimum time since death in indoor cases. J Forensic Sci 60(2):525–531

Byrd JH, Castner JL (2001) Insects of forensic importance. Forensic entomology: the utility of arthropods in legal investigations, vol 418. CRC Press, Boca Raton, FL, pp 43–79

Byrd JH, Castner JL (2010) Forensic entomology the utility of arthropods in legal investigations. Taylor & Francis, Boca Raton, FL

Campobasso CP, Linville JG, Wells JD, Introna F (2005) Forensic genetic analysis of insect gut contents. Am J Forensic Med Pathol 26(2):161–165

Chakraborty A, Ansar W, Ghosh A, Banerjee D (2014) The first report of the life cycle of Sarcophaga (L) dux on dead reptilian carcass: their application as forensic indicators. Sch Acad J Biosci 2(11):731–739

Cossé AA, Baker TC (1996) House flies and pig manure volatiles: wind tunnel behavioral studies and electrophysiological evaluations. J Agric Urban Entomol 13:301–317

Cragg JB, Ramage GR (1945) Chemotropic studies on the blowflies Lucilia sericata (mg.) and Lucilia caesar (L.). Parasitology 36:168–175

Cragg JB, Thurston BA (1950) The reactions of blowflies to organic Sulphur compounds and other materials used in traps. Parasitology 40:187–194

Crook A (2007) Forensic investigation of animal cruelty: a guide for veterinary and law enforcement professionals. Can Vet J 48(9):941–942

Dadour IR, Cook DF, Fissioli JN, Bailey WJ (2001) Forensic entomology: application, education and research in Western Australia. Forensic Sci Int 120(1–2):48–52

Defilippo F, Rubini S, Dottori M, Bonilauri P (2016) The use of forensic entomology in legal veterinary medicine: a case study in the north of Italy. J Forensic Sci Criminol 4(1):4–8

Dekeirsschieter J, Frederickx C, Lognay G, Brostaux Y, Verheggen FJ, Haubruge E (2013) Electrophysiological and behavioral responses of thanatophilus sinuatus fabricius (Coleoptera: Silphidae) to selected cadaveric volatile organic compounds. J Forensic Sci 58:917–923

Forensic Working Group (2014) Wildlife Crime A guide to the use of forensic and specialist techniques in the investigation of wildlife crime, Produced 2014 by the Forensic Working Group (FWG), part of the Partnership for Action against Wildlife Crime (PAW)

Frederickx C, Dekeirsschieter J, Verheggen FJ, Haubruge E (2012) Responses of Lucilia sericata Meigen (Diptera: Calliphoridae) to cadaveric volatile organic compounds. J Forensic Sci 57: 386–390

Frederickx C, Verheggen FJ, Brostaux Y, Haubruge E (2014) Associative learning of Nasonia vitripennis Walker (hymenoptera: Pteromalidae) to methyldisulfanylmethane. J Forensic Sci 59: 413–416

Gujar MG, Chouhan AK, Sodha DB, Rathore B (2020) Forensic entomology: insects as evidence. J Entomol Zool Stud 8(6):2059–2061

Guo Y, Cai J, Chang Y, Li X, Liu Q, Wang X, Wang X, Zhong M, Wen J, Wang J (2011) Identification of forensically important sarcophagid flies (Diptera: Sarcophagidae) in China, based on COI and 16S rDNA gene sequences. J Forensic Sci 56(6):1534–1540

Huffman JE, Wallace JR (2012) Wildlife forensics: methods and applications. John Wiley & Sons, New York, pp 82–83

Introna F, Campobasso CP, Goff ML (2001) Entomotoxicology. Forensic Sci Int 120(1–2):42–47

Johansen H, Solum M, Knudsen GK, Hågvar EB, Norli HR, Aak A (2014) Blow fly responses to semiochemicals produced by decaying carcasses. Med Vet Entomol 28:26–34

Kalinová B, Podskalská H, Růzicka J, Hoskovec M (2009) Irresistible bouquet of death: how are burying beetles (Coleoptera: Silphidae: Nicrophorus) attracted by carcasses? Naturwissenschaften 96:889–899

Kotzé Z, Aimar S, Amendt J, Anderson GS, Bourguignon L, Hall MJR, Tomberlin JK (2021) The forensic entomology case report-a global perspective. Insects 12(4):283

Lei G, Liu F, Liu P, Zhou Y, Jiao T, Dang YH (2019) A bibliometric analysis of forensic entomology trends and perspectives worldwide over the last two decades (1998-2017). Forensic Sci Int 295:72–82

McGarry J, Ratsep E, Ressel L, Leeming G, Ricci E, Verin R, Blundell R, Kipar A, Hetzel U, Yeates J (2018) Introducing forensic entomology in cases of suspected animal neglect. Vet Rec 182(5):139

Meiklejohn KA, Wallman JF, Dowton M (2011) DNA-based identification of forensically important Australian Sarcophagidae (Diptera). Int J Legal Med 125(1):27–32

Meiklejohn KA, Dowton M, Pape T, Wallman JF (2013) A revised key to the Australian Sarcophagidae (Diptera) with special emphasis on Sarcophaga (*sensulato*). Zootaxa 3680(1): 148–189

Nilssen AC, Tømmerås BÅ, Schmid R, Evensen SB (1996) Dimethyl trisulphide is a strong attractant for some calliphorids and a muscid but not for the reindeer oestrids Hypoderma tarandi and Cephenemyia trompe. Entomol Exp Appl 79:211–218

Podskalská H, Růžička J, Hoskovec M, Šálek M (2009) Use of infochemicals to attract carrion beetles into pitfall traps. Entomol Exp Appl 132:59–64

Ridgeway JA, Midgley JM, Collett IJ, Villet MH (2014) Advantages of using development models of the carrion beetles Thanatophilus micans (Fabricius) and T. mutilatus (Castelneau) (Coleoptera: Silphidae) for estimating minimum post mortem intervals, verified with case data. Int J Legal Med 128(1):207–220

Rolo EA, Oliveira AR, Dourado CG, Farinha A, Rebelo MT, Dias D (2013) Identification of sarcosaprophagous Diptera species through DNA barcoding in wildlife forensics. Forensic Sci Int 228(1–3):160–164

Smith KGV (1986) A manual of forensic entomology. Cornell University Press, Ithaca, NY

Stensmyr MC, Urru I, Collu I, Celander M, Hansson BS, Angioy AM (2002) Pollination: rotting smell of dead-horse arum florets. Nature 420:625–626

Sukontason K, Bunchu N, Chaiwong T, Moophayak K, Sukontason KL (2010) Forensically important flesh fly species in Thailand: morphology and developmental rate. Parasitol Res 106(5):1055–1064

Tuccia F, Zurgani E, Bortolini S, Vanin S (2019) Experimental evaluation on the applicability of necrobiome analysis in forensic veterinary science. Microbiology 8(9):e00828

Verheggen F, Perrault KA, Megido RC, Dubois LM, Francis F, Haubruge E, Forbes SL, Focant JF, Stefanuto PH (2017) The odor of death: an overview of current knowledge on characterization and applications. Bioscience 67(7):600–613

Von Hoermann C, Ruther J, Ayasse M (2012) The attraction of virgin female hide beetles (Dermestes maculatus) to cadavers by a combination of decomposition odour and male sex pheromones. Front Zool 9:18

Williams KA, Villet MH (2006) A history of southern African research relevant to forensic entomology. SAf J Sci 102(1/2):59–65

Zhu JJ, Chaudhury MF, Tangtrakulwanich K, Skoda SR (2013) Identification of oviposition attractants of the secondary screwworm, Cochliomyia macellaria (F.) released from rotten chicken liver. J Chem Ecol 39:1407–1414

Climate Change and Forensic Flies

8

Abstract

Climate change is a global phenomenon affecting the plant, animal, and human life. Insect diversity contributes to ecosystem functions and services as it is vital to the ecosystem and climate change has been known to impact the invertebrates and insects causing stress. Thus, it is important to study the impact of climate change on insect population. Although few studies reports are there to conclusively understand the impact of climate change on invertebrate life forms and in insects, in this chapter we discuss the current knowledge regarding impact of climate change on biodiversity, distribution, physiology, community, structure, and its interactions in insects. We also outline the few studies reporting the impact of climate change on forensically important insects. It remains to be important to the policy makers on climate change while framing policies, to mitigate the loss of biodiversity and delivery of ecosystem services by insects.

Keywords

Insects · Biodiversity · Climate change · Forensic entomology

Abbreviations

AR	Assessment Report
CE	Climate extreme
CH_4	Methane
CO_2	Carbon dioxide
D. semiclausum	*Diadegma semiclausum*
G. texensis	*Gryllus texensis*
GHG	Greenhouse gases
GMST	Global Mean Surface Temperature

S. Ghosh, D. Banerjee, *Biology of Forensically Important Invertebrates*,
https://doi.org/10.1007/978-981-99-5026-3_8

IPCC	Intergovernmental Panel on Climate Change
L	Light intensity
L. cuprina	*Lucilia cuprina*
L. sericata	*Lucilia sericata*
N_2O	Nitrous oxide
NO_2	Nitrogen dioxide
O_3	Ozone
R	Rainfall
RH	Relative Humidity
SO_2	Sulfur dioxide
T	Temperature
TE	Temperature extremes
VOC	Volatile organic compounds
W	Wind velocity

8.1 Introduction: Climate Change—A Global Perspective

The climate of the earth is changing, due to the predominant role played by anthropogenic activities. The addition of the greenhouse gases (GHG) in the atmosphere that can absorb and re-emit infrared (IR) radiations from the atmosphere down to the earth's surface, thereby trapping heat and causing greenhouse effect, leads to increased earth's mean surface temperature, reflected from the heights of the atmosphere to the oceans. These gases include carbon dioxide (CO_2), methane (CH4), nitrous oxide (N_2O), halocarbons and sulfur hexafluoride (SF_6), water vapor, molecular hydrogen (H_2), and methane (CH_4) (Table 8.1).

Since the nineteenth century, human activities like fossil fuels, deforestation, and land use changes have caused increased atmospheric concentrations of GHG, like CO_2, CH_4, N_2O, and fluorinated gases of which CO_2 that trap heat and form the main long-lived atmospheric GHG generated from anthropogenic sources.

The United Nations Intergovernmental Panel on Climate Change (IPCC) in their assessment reports reveals progressive and fast increase of atmospheric CO_2 from 280 parts/million to more than 416 parts/million during the 1850s to February 2020 and reveals the role of anthropogenic forces in causing global warming by about 1 °C during the last 150 years, leading to events of glacier melting, rise in sea levels,

Table 8.1 Main GHG whose concentrations are rising in lower atmosphere

Carbon dioxide (CO_2)
Methane (CH_4)
Nitrous oxide(N_2O)
Hydrochlorofluorocarbons (HCFCs)
Perfluorocarbons (PFC)
Hydrofluorocarbons (HFCs)
Ozone (O_3)

altered precipitation, and an increased weather and climate extremes. These impacts of climate changes are projected to continue through the twenty-first century, with the rise in GHG concentrations.

The IPCC Fifth Assessment Report (AR5) reported increased climate change response of the precipitation during Indian monsoons with major impact on rise in Indian Ocean sea level and melting of Himalayan snow cover, together with other impacts.

In the twenty-first century, climate change and its direct impact are conferring challenges to human health, safety, life quality, and economy globally. Although climate change is a global phenomenon, climate change lacks uniformity across the globe. The local climate change is influenced by increased GHG, air pollution, and local land-use pattern alterations.

The impacts of climate change with long-term alteration in temperatures and altered weather patterns are negative and are expected to become far more disruptive with years to come across the planet. Not only is animal life affected, but studies show that impact of climate change has largely impacted the animal, plant, and wildlife negatively. It is important for the policy makers to have a clear vision on the possible future projected climate changes and study of climate models. An assessment report of climate change reveals observed changes and future projections of increased temperature, precipitation, water scarcity, heavy monsoon, floods, major drought, severe fires, polar ice melting, catastrophic storms, tropical cyclones, extreme weather events, and declining biodiversity (Figs. 8.1 and 8.2).

8.2 Climate Change and Impact on Insects

Climate change is known to play an important role in the decline of biodiversity at a global level, altering freshwater, marine, and terrestrial ecosystems together with loss of local species, increase in diseases, mass death of plants and animals, and extinctions. On land, with the increase of temperatures, animals and plants have been observed to move up at higher elevations with some even moving to the poles. Each degree rise in global temperature is associated with extinction of life forms.

Climate change has impacted largely the human and wildlife globally. In this chapter we focus our attention to the impact of climate change and abiotic and biotic factors underlying these variations on the insect life (Adler and Courtney 2019; Anderegg et al. 2022; Arce-Valdés and Sánchez-Guillén 2022; Freimuth et al. 2022; Garnas 2018; Giersch et al. 2017; Popular Science (n.d.); Rafferty 2017; Diamond 2018; Raven and Wagner 2021; St Leger 2021).

Insects with their diversity contribute to ecosystem functions and services. Increased frequency and extent and nature of climatic extremes (CE) and temperature extremes (TE) owing to climate change and anthropogenic activities are known to cause stress to the insect populations (Harvey et al. 2020). Many insects on mountains, surving in steep and rapid climatic gradients, like most 1478 montane insect populations have been reported to shift their elevational range limits in response to recent alterations in temperature (McCain and Garfinkel 2021).

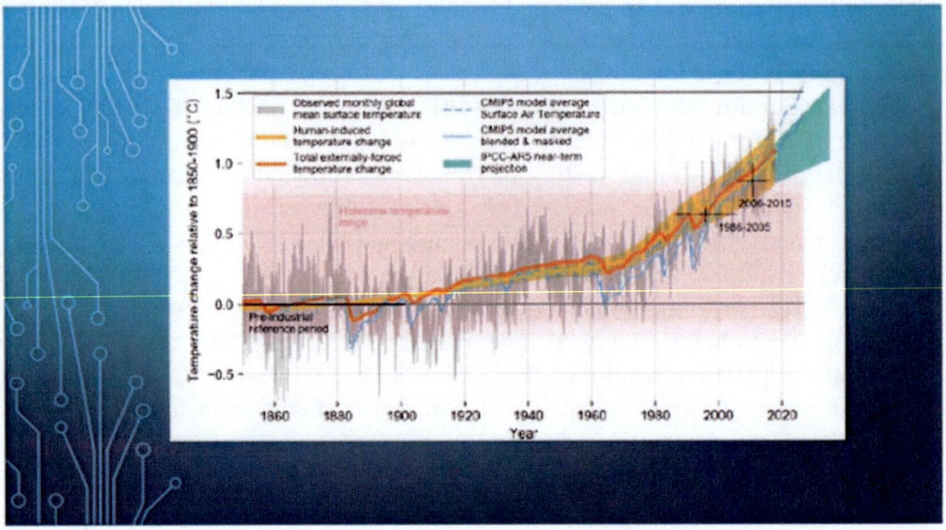

Fig. 8.1 Global Mean Surface Temperature (GMST) as weighted average of near-surface air temperature over land and sea surface temperature over ocean from 1850 to 1900. Human-induced contributions marked by the yellow line, total human and natural forces represented by the orange line. The blue dashed lines reveal modeled GMST, and solid lines represent surface air and sea surface temperature from CMIP5 historical perspective average extended with RCP.5. The pink shade indicates range of temperature fluctuations over the Holocene (Marcott et al. 2013). Light green plume shows AR5 prediction for average GMST over 2016–2035 (Kirtman et al. 2013; Allen et al. 2018; Otto et al. 2015; Haustein et al. 2017; Cowtan et al. 2015; Richardson et al. 2018. Figure and legend adapted from open-access book Krishnan et al. 2020)

Insect diversity dates back far greater than 450×10^6 years of earth's climate, but rapid shifts of temperature and precipitation impacts on insects like butterflies of Northern California cause severe decline in population in high-elevation areas to being removed due to habitat loss (Halsch et al. 2021), thereby threatening the biodiversity.

Climate change and extreme weather conditions are known to impact plants and animals negatively. Increase in CO_2 and O_3 concentrations, increased surface temperature, and altered rainfall patterns impact insect-host plant interactions (Cornelissen 2011). Climate change-mediated multitrophic interaction reveals dramatic shift in insect distribution together with their pathogens, forcing some populations toward extinction. Both virulence of pathogen and immunity of insects are affected by stress responses, affecting nutrition, growth, and reproduction (Ledger 2021). Climate change together with anthropogenic changes has been reported to cause reorganization of communities (Yang et al. 2021).

Populations and species have revealed a sharp decline in abundance and diversity due to impact of climate change. It is known to alter biotic interactions and ecosystem functions. It is also known to affect invasive social insects and alter their distribution and phenologies involving altered timing period of cyclical or seasonal events like hibernation, egg laying, migrations, flowering, changes in distribution ranges, and abundance (Menzel and Feldmeyer 2021).

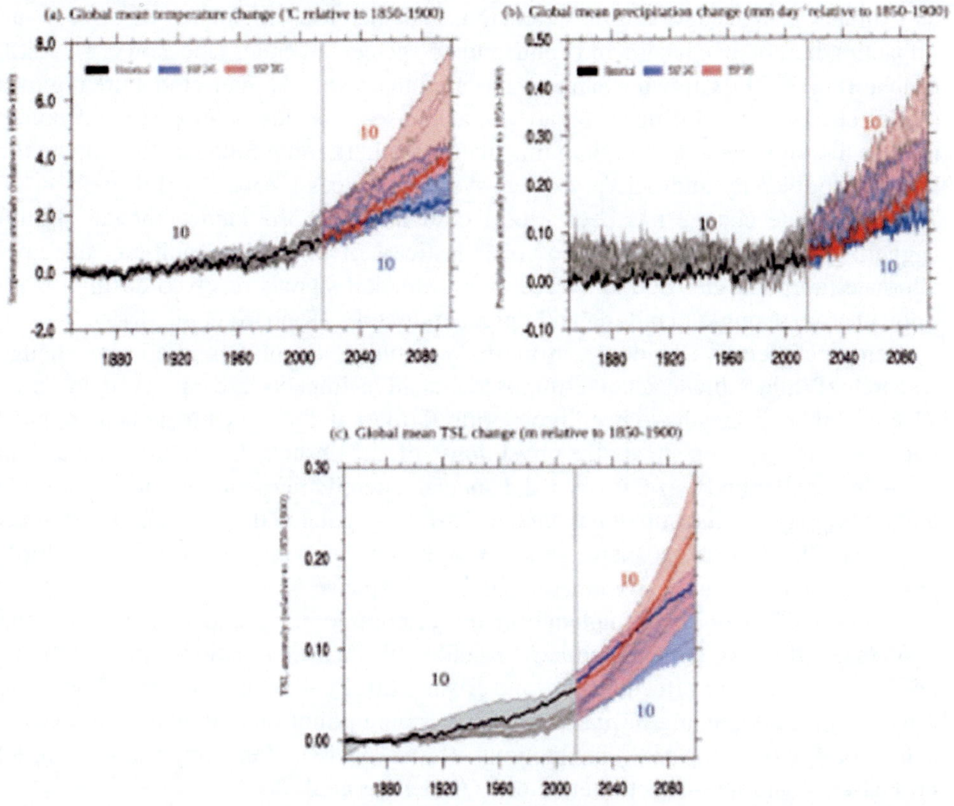

Fig. 8.2 Time series of CMIP6 simulations of the globally averaged annual mean. (**a**) Surface air temperature, (**b**) precipitation (mm/day), (**c**) thermosteric sea level (TCL: m). Plotted are anomalies with respect to the pre-industrial baseline (1850–1900) for the historical period (black) and future projections following the SSP2–4.5 (blue) and SSP5–8.5 (red) scenarios. Numbers represent number of models selected for each variable. Solid lines represent the IITM-ESM simulations, and the shading represents the range across the models (figure and legend adapted from open-access book Krishnan et al. 2020)

Kairomones are semiochemical allelochemicals that are emitted by insects and can attract insect predators enabling insect detection by predator, thereby causing harm to the emitting insect. Climate change and alterations in the atmospheric gases might impact the chemical communication between phytophagous insect pests, their host plants, and the natural predator species. Reports have indicated elevated CO_2 and O_3 concentrations leading to modifications to plants that might indirectly affect insect herbivores. Under high GHG concentrations, agricultural insect pests have been reported to be vulnerable to their natural enemies.

Increased O_3 levels could add to stress and cause modifications in damaged plant released herbivore-induced volatile organic compounds (VOCs). This could increasingly attract the herbivores to the higher trophic level. SO_2 and NO_2 are reported to decrease the natural enemy in abundance. Atmospheric gas emissions have been known to impact chemical communication between plant-insect and insect-insect (Boullis et al. 2015).

Climate change, agricultural extension, and increased incidence of disease are impacting bumblebee health and contribute to species' declines (Jackson et al. 2022; Maebe et al. 2021). Climate change is reported to drastically alter global fire regimes (Koltz et al. 2018) and pose threat to many insect species. Biologists and policy makers therefore need to look into this matter while framing policies, to mitigate the loss of biodiversity and delivery of ecosystem services by insects (Filazzola et al. 2021). Climate change has been reported to modulate the environmental stimuli triggering polyphenisms or phenotypic plasticity of producing multiple, different alternate morphologies due to altered environmental signals received during development from a single genotype and epigenetic marks (Richard et al. 2019).

Being ectoderms, Lepidopteran moths are studied as models in climate change research. While climate change-imposed negative impacts are observed in some species, other species have been prospering (Hill et al. 2021). Climate change have been reported to impact at the physiology of the insects including decrease in melanin, smaller body size, increased dispersal, altered microhabitats, across altitude and latitude, generalisation of habitat and host, up-regulated thermoregulation events like heat shock protein (hsp) genes expression, tendency of passing multiple generations/annum and early emergence from diapause.

Overwintering poses a challenge in insect communities, and with fast altering winters as effects of climate shifts (Marshall et al. 2020), ant colonies, populations, and communities are affected (Parr and Bishop 2022). Climate change is known to affect chemical signaling of plants and insects, and pollution and other components can modify predator-prey interactions (Laws 2017), tree physiology, insect populations, and tree-insect interactions (Anderegg et al. 2015).

Although insects are highly successful in regulating their body temperature, climate change together with rise in global temperature is reported to alter the range and distribution of stinging insects. Many species have been reported to expand toward the poles, due to global warming (Demain 2021), and are undergoing or are projected to undergo shifts in their geographic ranges, voltinism, abundance, population dynamics, and phenology (Eigenbrode et al. 2022; Boggs 2016). Climate change has known to facilitate biological invasions of hymenopteran species (Vega and Castro 2019), impact phenology of many species of both plants and insects, and impact plant-pollinator interactions (Kingsolver et al. 2011).

Studies on dependence on temperature for growth of insect population have revealed that altered temperature can impact juvenile survival and adult reproduction in different ways (Kingsolver et al. 2011).

In response to increased stress due to climate change, growth rates in insect population are predicted to increase in the temperate zone. Texas field cricket population of *Gryllus texensis* first described by William H. Cade and Otte in the year 2000 revealed decline in numbers with increased surface temperature, reduced availability of food, and decreased survival simulated by heat wave (Adamo et al. 2012).

Climate change is globally altering conditions in streams in high-elevated areas, but the effects on resident aquatic insect communities are largely unknown. While short-term responses to climate change may be governed by tolerances of individuals, longer-term responses may include impact on individual plasticity and population evolution. But detailed study on mechanism on aquatic resident insects at high elevation that can mitigate risks associated with climatic risks remains largely unknown (Birrell et al. 2020).

Global climate change has led to shifting of species range to higher latitudes. It has impacted on invertebrates. Climate change is also contributing to reduced biodiversity and global insect loss (Sirois-Delisle and Kerr 2022).

Insect communication by use of pheromones, encompassing their behavioral response, is predicted to be impacted by increased temperature and altered atmospheric CO_2 and O_3 levels. Thus, insect species that depends on long-range chemical signals are known to be mostly impacted due to the fact that these signals are thought to be affected more due to prolonged exposure to oxidative gases while dispersal (Boullis et al. 2016).

Due to the correlation between climate change and agriculture, fast developing climate change has impacted agroecosystems and productivity of crops. Rising temperatures and CO_2 levels have been reported to affect insect physiology and increase metabolism and consumption, leading to increased population densities, causing greater injury and damage to crops and vegetation, and yielding economic and agricultural losses (Tonnang et al. 2022). These impacts of climate change on agriculture are hypothesized to alter pest pressures on plant communities like weeds like kudzu and insects (like mountain pine beetle) which have undergone expansion in northern regions due to increased temperature (Ziska and McConnell 2016).

Studies on mountain butterflies and their distribution across the eastern Alps revealed climatic conditions and variable topography, of isolation and solar irradiation as the main drivers (Rödder et al. 2021) for their long-term distribution.

Incidence of vector-borne animal diseases is threatening to livestock industries and economies globally. Many climatic variables impact the distribution and range of insects like mosquitoes, *Culicoides*. While increase in temperature is associated with increased abundance of insects with increased disease transmission, no studies have been conducted to study shifts in geographic range of *Culicoides*. Precipitation and extreme conditions like severe droughts and floods may affect mosquito dispersal (Elbers et al. 2015).

Although insects are ectotherm organisms, summer heat and drought impact on insect diversity are not well studied. In the Mediterranean climate zone, hot summers may impact insect communities severely, with sensitive larval stage to thermal variation indicating that severity of increased temperature and drought extremes due to climate change might cause decline in insects (Uhl et al. 2022).

Hymenopteran parasitoids require optimal temperature for their development which was less than that required for their hosts. Interactions between the diamondback moth and its parasitoid *Diadegma semiclausum* revealed that predicted temperature increases might impact negatively the parasitoid distribution than that of the host leading to their exclusion from some required agricultural zones (Furlong and Zalucki 2017).

8.3 Climate Change and Impact on Forensic Insects

Forensic entomology is usage of insects, their identification, and tables of insect succession in order to estimate the PMI. In the last few years, studies on climate change revealed alteration in faunal communities by favoring thermophilous species of southern Italy and allochthonous species of north-east Italy in forensic entomological investigations (Turchetto and Vanin 2004).

Recent studies on climate change and impact on insects reveal loss in insect species, biodiversity, abundance, and biomass. A decreased species diversity and population density in forensically important insects could lead to a delayed colonization of dead bodies. This could also lead to altered insect succession pattern due to the disappearance or emergence of new species or altered seasonality. Climate change is thought to be an important drivers of insect decline and impact forensic entomology. The impact could cause reduced flight, reduced reproduction/oviposition activity, and modification in growth rates which in turn may lead to inaccuracy in estimation of PMI_{min} (over- or underestimation).

Global warming with extreme weather conditions and increased temperature needs to be studied more to understand the impact in the context of necrophagous insects and to understand their responses to environmental variations. Effects of transgeneration may play role in fast adaptation and plasticity in forensically important importance (Amendt 2021).

Rise in temperatures globally is enforcing migration of blow flies to northern areas in search of cooler weather, affecting the crime investigations used by forensic scientists.

Lucilia cuprina, or Australian sheep blow fly, has been reported from Midwest, southern states , reported in parks and other public places in Indiana between 2015 and 2017 bearing similarity to sister species, *Lucilia sericata*, or greenbottle fly widely distributed in the state. However, the maggots reveal differences. It is predicted that with increased temperature changes, the insect distribution will alter as more migrations northward in search of cooler climates and new food sources are observed (https://www.technologynetworks.com/applied-sciences/news/knock-on-of-climate-change-for-forensics-experts-311318, Owings and Picard 2018).

Studies on climate change and its impacts on forensic insects are few and require more analysis and observations as global warming may alter the PMI estimation accuracy if forensic insects growing on body are not carefully studied. Time disparity in PMI estimation of a body is thought to be impacted due to climate change in the tropics and globally. Climate change is hypothesized to incontrovertibly alter the biological clock and distribution range of forensically important insects.

In Nigeria, studies of climate change on forensically important insects reveal early arrival and shorter life cycle of forensic insects with faster decomposition of cadavers in the southern region. Climatic variables of relative humidity, temperature, precipitation, rainfall, wind velocity, and light intensity have been reported to impact insect population dynamics by modulating their distribution range, abundance, survival, life cycle, biological clocks, development rate, and fecundity (Maduamaka et al. 2020). In a study by the National Geographic, the forensic insects have been reported to shift northward (Jennifer 2013).

Christine Picard, at Indiana University-Purdue University Indianapolis, discovered *Chrysomya megacephala*, native to the southern USA and reported from Indiana, north of New Mexico, indicative of consequences of global climate change.

8.4 Discussion

Thus, although only a few studies have been conducted in the detection and conclusively report on the impact of climate change on forensic insects, there is a clear indication from other studies on insects and impact of climate change on them that the biomass, distribution pattern, reproduction, growth, time of colonization, life cycle, physiology, etc. may be affected. However, at the same time, this knowledge gap calls in for more research in this area as it might affect the whole investigation method based on altered colonization, life cycle, and altered succession pattern in the dead bodies.

References

Adamo SA, Baker JL, Lovett MM, Wilson G (2012) Climate change and temperate zone insects: the tyranny of thermodynamics meets the world of limited resources. Environ Entomol 41(6): 1644–1652

Adler PH, Courtney GW (2019) Ecological and societal Services of Aquatic Diptera. Insects 10(3): 70

Allen MR, Dube OP, Solecki W, Aragón-Durand F, Cramer W, Humphreys S, Kainuma M, Kala J, Mahowald N, Mulugetta Y, Perez R, Wairiu M, Zickfeld K (2018) Framing and context. In: Masson-Delmotte V, Zhai P, Pörtner H-O, Roberts D, Skea J, Shukla PR, Pirani A, Moufouma-Okia W, Péan C, Pidcock R, Connors S, Matthews JBR, Chen Y, Zhou X, Gomis MI, Lonnoy E, Maycock T, Tignor M, Waterfield T (eds) Global warming of 1.5°C. An IPCC Special Report on the impacts of global warming of 1.5°C above pre-industrial levels and related global GHG emissions pathways, in the context of strengthening the global response to the threat of climate change, sustainable development, and efforts to eradicate poverty

Amendt J (2021) Insect decline—a Forensic issue? Insects 12(4):324. https://doi.org/10.3390/insects12040324

Anderegg WR, Hicke JA, Fisher RA, Allen CD, Aukema J, Bentz B, Hood S, Lichstein JW, Macalady AK, McDowell N, Pan Y, Raffa K, Sala A, Shaw JD, Stephenson NL, Tague C, Zeppel M (2015) Tree mortality from drought, insects, and their interactions in a changing climate. New Phytol 208(3):674–683

Anderegg WRL, Chegwidden OS, Badgley G, Trugman AT, Cullenward D, Abatzoglou JT, Hicke JA, Freeman J, Hamman JJ (2022) Future climate risks from stress, insects and fire across US forests. Ecol Lett 25(6):1510–1520

Arce-Valdés LR, Sánchez-Guillén RA (2022) The evolutionary outcomes of climate-change-induced hybridization in insect populations. Curr Opin Insect Sci 54:100966

Birrell JH, Shah AA, Hotaling S, Giersch JJ, Williamson CE, Jacobsen D, Woods HA (2020) Insects in high-elevation streams: life in extreme environments imperiled by climate change. Glob Chang Biol 26(12):6667–6684

Boggs CL (2016) The fingerprints of global climate change on insect populations. Curr Opin Insect Sci 17:69–73

Boullis A, Francis F, Verheggen FJ (2015) Climate change and Tritrophic interactions: will modifications to greenhouse gas emissions increase the vulnerability of herbivorous insects to natural enemies? Environ Entomol 44(2):277–286

Boullis A, Detrain C, Francis F, Verheggen FJ (2016) Will climate change affect insect pheromonal communication? Curr Opin Insect Sci 17:87–91

Cornelissen T (2011) Climate change and its effects on terrestrial insects and herbivory patterns. Neotrop Entomol 40(2):155–163

Cowtan K, Hausfather Z, Hawkins E, Jacobs P, Mann ME, Miller SK, Steinman BA, Stolpe MB, Way RG (2015) Robust comparison of climate models with observations using blended land air and ocean sea surface temperatures. Geophys Res Lett 42:6526–6534

Demain JG (2021) Insect migration and changes in venom allergy due to climate change. Immunol Allergy Clin N Am 41(1):85–95

Diamond SE (2018) Contemporary climate-driven range shifts: putting evolution back on the table. Funct Ecol 32:1652–1665. https://doi.org/10.1111/1365-2435.13095

Eigenbrode SD, Adhikari S, Kistner-Thomas E, Neven L (2022) Introduction to the collection: climate change, insect pests, and beneficial arthropods in production systems. J Econ Entomol 115(5):1315–1319

Elbers AR, Koenraadt CJ, Meiswinkel R (2015) Mosquitoes and culicoides biting midges: vector range and the influence of climate change. Rev Sci Tech 34(1):123–137

Filazzola A, Matter SF, MacIvor JS (2021) The direct and indirect effects of extreme climate events on insects. Sci Total Environ 769:145161

Freimuth J, Bossdorf O, Scheepens JF, Willems FM (2022) Climate warming changes synchrony of plants and pollinators. Proc Biol Sci 289(1971):20212142

Furlong MJ, Zalucki MP (2017) Climate change and biological control: the consequences of increasing temperatures on host-parasitoid interactions. Curr Opin Insect Sci 20:39–44

Garnas JR (2018) Rapid evolution of insects to global environmental change: conceptual issues and empirical gaps. Curr Opin Insect Sci 29:93–101

Giersch JJ, Hotaling S, Kovach RP, Jones LA, Muhlfeld CC (2017) Climate-induced glacier and snow loss imperils alpine stream insects. Glob Chang Biol 23(7):2577–2589

Halsch CA, Shapiro AM, Fordyce JA, Nice CC, Thorne JH, Waetjen DP, Forister ML (2021) Insects and recent climate change. Proc Natl Acad Sci U S A 118(2):e2002543117

Harvey JA, Heinen R, Gols R, Thakur MP (2020) Climate change-mediated temperature extremes and insects: from outbreaks to breakdowns. Glob Chang Biol 26(12):6685–6701

Haustein K, Allen MR, Forster PM, Otto FEL, Mitchell DM, Matthews HD, Frame DJ (2017) A real-time global warming index. Sci Rep 7:15417

Hill GM, Kawahara AY, Daniels JC, Bateman CC, Scheffers BR (2021) Climate change effects on animal ecology: butterflies and moths as a case study. Biol Rev Camb Philos Soc 96(5): 2113–2126

Jackson HM, Johnson SA, Morandin LA, Richardson LL, Guzman LM, M'Gonigle LK (2022) Climate change winners and losers among North American bumblebees. Biol Lett 18(6): 20210551

Jennifer SH (2013) Forensic fly moves north, insects can aid death investigations, but non-native species complicate matters. Natl Geogr

Kingsolver JG, Woods HA, Buckley LB, Potter KA, MacLean HJ, Higgins JK (2011) Complex life cycles and the responses of insects to climate change. Integr Comp Biol 51(5):719–732

Kirtman B et al (2013) Near-term climate change: projections and predictability. In: Stocker TF, Qin D, Plattner G-K, Tignor M, Allen SK, Boschung J, Nauels A, Xia Y, Bex V, Midgley PM (eds) Climate change 2013: the physical science basis. Contribution of working group I to the fifth assessment report of the intergovernmental panel on climate change. Cambridge University Press, Cambridge, pp 953–1028

Koltz AM, Burkle LA, Pressler Y, Dell JE, Vidal MC, Richards LA, Murphy SM (2018) Global change and the importance of fire for the ecology and evolution of insects. Curr Opin Insect Sci 29:110–116

Krishnan R, Sanjay J, Gnanaseelan C, Mujumdar M, Kulkarni A, Chakraborty S (2020) Assessment of Climate Change over the Indian Region A Report of the Ministry of Earth Sciences (MoES), Government of India. Springer, Singapore. https://doi.org/10.1007/978-981-15-4327-2

Laws AN (2017) Climate change effects on predator-prey interactions. Curr Opin Insect Sci 23:28–34

Maduamaka CA, Nwosu LC, Ileke KD (2020) Implications of global changing climate on forensic insects in the tropics: a review. J Entomol Zool Stud 8(5):1713–1719

Maebe K, Hart AF, Marshall L, Vandamme P, Vereecken NJ, Michez D, Smagghe G (2021) Bumblebee resilience to climate change, through plastic and adaptive responses. Glob Chang Biol 27(18):4223–4237

Marcott SA, Shakun JD, Clark PU, Mix AC (2013) A reconstruction of regional and global temperature for the past 11,300 years. Science 339(6124):1198–1201

Marshall KE, Gotthard K, Williams CM (2020) Evolutionary impacts of winter climate change on insects. Curr Opin Insect Sci 41:54–62

McCain CM, Garfinkel CF (2021) Climate change and elevational range shifts in insects. Curr Opin Insect Sci 47:111–118

Menzel F, Feldmeyer B (2021) How does climate change affect social insects? Curr Opin Insect Sci 46:10–15

Otto FEL, Frame DJ, Otto A, Allen MR (2015) Embracing uncertainty in climate change policy. Nat Clim Chang 5:1–5. https://doi.org/10.1038/nclimate2716

Owings CG, Picard CJ (2018) New distribution record for Lucilia cuprina (Diptera: Calliphoridae) in Indiana, United States. J Insect Sci 18(4):8. https://doi.org/10.1093/jisesa/iey071

Parr CL, Bishop TR (2022) The response of ants to climate change. Glob Chang Biol 28(10): 3188–3205

Popular Science (n.d.). https://www.popsci.com/crime-solving-insects-climate-change/ing temperatures are driving these crime-fighting insects from their homes

Rafferty NE (2017) Effects of global change on insect pollinators: multiple drivers lead to novel communities. Curr Opin Insect Sci 23:22–27. https://doi.org/10.1016/j.cois.2017.06.009. Epub 2017 Jul 1. PMID: 29129278

Raven PH, Wagner DL (2021) Agricultural intensification and climate change are rapidly decreasing insect biodiversity. Proc Natl Acad Sci U S A 118(2):e2002548117

Richard G, Le Trionnaire G, Danchin E, Sentis A (2019) Epigenetics and insect polyphenism: mechanisms and climate change impacts. Curr Opin Insect Sci 35:138–145

Richardson M, Cowtan K, Millar RJ (2018) Global temperature definition affects achievement of long-term climate goals. Environ Res Lett 13(5):054004

Rödder D, Schmitt T, Gros P, Ulrich W, Habel JC (2021) Climate change drives mountain butterflies towards the summits. Sci Rep 11(1):14382

Sirois-Delisle C, Kerr JT (2022) Climate change aggravates non-target effects of pesticides on dragonflies at macroecological scales. Ecol Appl 32(2):e2494

Leger RJ (2021) Insects and their pathogens in a changing climate. J Invertebr Pathol 184:107644

Tonnang HE, Sokame BM, Abdel-Rahman EM, Dubois T (2022) Measuring and modelling crop yield losses due to invasive insect pests under climate change. Curr Opin Insect Sci 50:100873

Turchetto M, Vanin S (2004) Forensic entomology and climatic change. Forensic Sci Int 146 (Suppl):S207–S209

Uhl B, Wölfling M, Bässler C (2022) Mediterranean moth diversity is sensitive to increasing temperatures and drought under climate change. Sci Rep 12(1):14473

Vega A, Castro L (2019) Impact of climate change on insect-human interactions. Curr Opin Allergy Clin Immunol 19(5):475–481

Yang LH, Postema EG, Hayes TE, Lippey MK, MacArthur-Waltz DJ (2021) The complexity of global change and its effects on insects. Curr Opin Insect Sci 47:90–102

Ziska LH, McConnell LL (2016) Climate change, carbon dioxide, and Pest biology: monitor, mitigate. Manage J Agric Food Chem 64(1):6–12